# science moves 2

Anne Garton

## Acknowledgments

The author and publisher would like to thank the following for granting their permission to reproduce copyright material:

Cooee Photo Library, p. 161 top, centre; Malcolm Cross, pp. 5 top, centre, 6, 32, 35 right, 50, 51, 60, 128, 129 top, bottom left, right, 132; Mark Fergus, pp. 30, 42 bottom left, 57, 83 top left, 169 centre, right, 170 right; 'Mousetrap' © 1996, Milton Bradley Company, all rights reserved, used with permission, p. 126; The Photo Library, pp. 10, 35 left, 40 left, 168 left, 174 both; Planet Earth, p. 40 right; Retrospect/Dale Mann, pp. 8, 19, 28, 61, 134; Bill Thomas, p. 83 bottom; Lyn Trounson, pp. 168 right, 169 left.

Every effort has been made to trace and acknowledge copyright material. The author and publisher would welcome any information from people who believe they own copyright material in this book.

Rigby Heinemann
a division of Reed International Books Australia Pty Ltd
22 Salmon Street, Port Melbourne, Victoria 3207
World Wide Web   http://www.reedbooks.com.au
Email   heinemann@reedbooks.com.au

Offices in Sydney, Brisbane, Adelaide and Perth.
Associated companies, branches and representatives throughout the world.

© Rigby Heinemann 1996
First published 1996
This edition contains some material originally published by John Murray (Publishers) Ltd in the *Understanding Science* series (by Joe Boyd and Walter Whitelaw) and adapted by Bernadette Tolan for the *Heinemann Science in Context* series, published by Rigby Heinemann.

All rights reserved. No part of this publication may be reproduced, stored in a retrieval system or transmitted in any form or by any means whatsoever without the prior permission of the copyright owner. Apply in writing to the publisher.

Edited by Stephen Dobney
Designed by Patricia Tsiatsias
Technical illustrations by Guy Holt and Brent Hagen
Freehand illustrations by Andrew Plant and Craig Smith
Cover design by Marion Marks
Cover photograph © The Image Bank
Photograph research by Jan Calderwood

Typeset in Berkeley and Gill Sans by Palmer Higgs Pty Ltd
Printed in Australia by Impact Printing

National Library of Australia
cataloguing-in-publication data:

Garton, Anne.

   ScienceMoves 2.
   Includes index.
   ISBN 0 85859 880 9.

   1. Science—Juvenile literature. I. Title.

500

# contents

| | | |
|---|---|---|
| How to use this book | | v |
| Making a glossary | | vi |
| CSF outcome statement | | vii |
| National Profile outcome statement | | ix |

## Working scientifically
### Section one
### Observing, recording and hypotheses

| Chapter 1 | Let's experiment | 2 |
|---|---|---|
| 1.1 | Hypothesis-making and scientific method | 3 |
| 1.2 | Spot the variables | 5 |
| 1.3 | Fair experiments | 6 |
| 1.4 | Recording the experiment | 7 |
| 1.5 | Project: Let's experiment | 9 |
| 1.6 | Science in action: The Leaning Tower of Pisa | 10 |
| 1.7 | Another look | 11 |
| Skillbuilder | Making a bibliography | 13 |

| Chapter 2 | Sun sense | 14 |
|---|---|---|
| 2.1 | The skin | 15 |
| 2.2 | It's great to see you | 17 |
| 2.3 | It's a matter of taste | 20 |
| 2.4 | Project: I hear you | 21 |
| 2.5 | Science in action: Can you tan safely? | 22 |
| 2.6 | Another look | 24 |
| Skillbuilder | Reviewing your work | 26 |

## Natural and processed materials
### Section two
### Elements and compounds
### Science in context

| Chapter 3 | Elements around us | 28 |
|---|---|---|
| 3.1 | Meet the elements | 29 |
| 3.2 | Where do you find the elements? | 30 |
| 3.3 | Introducing the metals | 32 |
| 3.4 | How metals are used | 34 |
| 3.5 | Metals and non-metals | 35 |
| 3.6 | Science in action: The Earth's resources | 36 |
| 3.7 | Another look | 37 |
| Skillbuilder | Writing skills: using fewer words in sentences | 39 |

| Chapter 4 | Elements make compounds | 40 |
|---|---|---|
| 4.1 | Atoms and molecules | 41 |
| 4.2 | Making compounds | 42 |
| 4.3 | Breaking compounds | 44 |
| 4.4 | Rust: an important compound | 45 |
| 4.5 | Chemical reactions | 47 |
| 4.6 | Numbers, atoms and formulae | 49 |
| 4.7 | Science in action: Wonderful chlorine | 50 |
| 4.8 | Project: Compounds in action | 51 |
| 4.9 | Another look | 52 |
| Skillbuilder | Reading skills: finding information | 54 |

| Chapter 5 | Acids and bases | 55 |
|---|---|---|
| 5.1 | Investigating acids and bases | 56 |
| 5.2 | Introducing indicators | 57 |
| 5.3 | Neutralisation | 59 |
| 5.4 | Science in action: The fire extinguisher | 60 |
| 5.5 | Acid problems | 61 |
| 5.6 | Project: Reaction riddles | 63 |
| 5.7 | Another look | 64 |
| Skillbuilder | Making your own criteria assessment sheets | 66 |

## Life and living
### Section three
### First aid, cells and the environment
### Science in context

| Chapter 6 | Life matters | 68 |
|---|---|---|
| 6.1 | What do I do first? | 69 |
| 6.2 | The lungs and breathing | 71 |
| 6.3 | How to save a life | 73 |
| 6.4 | The heart | 74 |
| 6.5 | Science in action: Divers and first aid | 78 |
| 6.6 | Another look | 79 |
| Skillbuilder | Understanding different diagrams | 81 |

| Chapter 7 | The bricks of life | 82 |
|---|---|---|
| 7.1 | The microscope | 83 |
| 7.2 | Let's look at cells | 86 |
| 7.3 | Processes inside cells | 88 |
| 7.4 | The jobs cells do | 90 |
| 7.5 | Science in action: Cancer | 92 |
| 7.6 | Another look | 93 |
| Skillbuilder | Reading skills: recognising fact and opinion | 95 |

| Chapter 8 | Environments | 96 |
|---|---|---|
| 8.1 | This is my habitat | 97 |
| 8.2 | Are you dinner? | 100 |
| 8.3 | Food webs | 103 |

| | | |
|---|---|---|
| 8.4 | All in balance | 104 |
| 8.5 | Science in action: Salty Australia | 106 |
| 8.6 | Investigating pollution | 107 |
| 8.7 | Project: Conservation awards | 108 |
| 8.8 | Another look | 110 |
| Skillbuilder | Writing skills: using diagrams in notes | 112 |

## Energy and change
### Section four
### Energy, machines and sound
### Science in context

| | | |
|---|---|---|
| **Chapter 9** | **Making things move** | **114** |
| 9.1 | Energy transformations | 115 |
| 9.2 | Conduction, convection and radiation | 118 |
| 9.3 | Project: Using what I've learned | 121 |
| 9.4 | Science in action: The solar cell | 122 |
| 9.5 | Another look | 123 |
| Skillbuilder | Writing skills: using diagrams to shorten notes | 125 |
| **Chapter 10** | **Moving machines** | **126** |
| 10.1 | What is a machine? | 127 |
| 10.2 | Different types of simple machines | 128 |
| 10.3 | Examining machines | 132 |
| 10.4 | Bigger machines | 133 |
| 10.5 | Science in action: The Internet | 134 |
| 10.6 | Another look | 135 |
| Skillbuilder | Writing skills: recording facts in notes | 137 |
| **Chapter 11** | **Sounds like …** | **138** |
| 11.1 | What is sound? | 139 |
| 11.2 | What does sound look like? | 140 |
| 11.3 | The frequency of sound waves | 142 |
| 11.4 | Sound and the ear | 143 |
| 11.5 | Other features of sound | 145 |
| 11.6 | What makes a note? | 147 |
| 11.7 | Turn that noise down | 149 |
| 11.8 | Science in action: Hearing loss | 151 |
| 11.9 | Another look | 152 |
| Skillbuilder | Mathematical calculations | 154 |

## The physical world / Earth and beyond
### Section five
### Rocks, earth and sky
### Science in context

| | | |
|---|---|---|
| **Chapter 12** | **Sun, moon and sky** | **156** |
| 12.1 | Aboriginal beliefs | 157 |
| 12.2 | Movement of the Earth and Moon | 158 |
| 12.3 | The heavens | 161 |
| 12.4 | Time zones | 163 |
| 12.5 | Science in action: The space shuttle | 164 |
| 12.6 | Another look | 165 |
| Skillbuilder | Using tables and data | 167 |
| **Chapter 13** | **Look beneath your feet** | **168** |
| 13.1 | Three types of rocks | 169 |
| 13.2 | The formation of rocks | 172 |
| 13.3 | Minerals and hardness | 174 |
| 13.4 | Rock for the road | 175 |
| 13.5 | Science in action: Coal | 177 |
| 13.6 | Another look | 178 |
| Skillbuilder | Reading and analytical skills | 180 |
| Index | | 181 |

# How to use this book

**To the student**

As you work through this book you will come across the following sections:

*Try this*

These are practical activities. They usually start with a 'Collect' list, which means that you will need to collect equipment or materials from your teacher to complete the activity. They will usually be followed by questions about the activity, and in some cases you will have to write up a formal practical report.

*Your turn*

These are general questions about the information presented in the text.

*On your own*

These are activities for you to work on with very little guidance. You may have to design your own experiment or carry out some further research.

*Science in action*

These are case studies that show you how science is used in the real world. There will usually be questions related to the case study, and sometimes you may be required to do further research.

*Project*

These occur in most chapters and may involve you drawing a diagram, building a model, making a poster or carrying out your own research. Your teacher may give you an assessment criteria sheet so that you know what is expected of you.

*Another look*

These are review sections that will help you revise the work you have done throughout the chapter.

*Skillbuilder*

These contain activities that will help you develop and practise important skills related to your work in science.

*New words*

At the end of each unit is a list of new words that have been introduced. You should add these words and their meanings to your own glossary of scientific terms. (See following page.)

**To the teacher**

Throughout this textbook you will see the following icons, which refer to material contained in the Teacher's Resource Pack.

These are references to the Teacher's Resource Book. They may refer to extension worksheets, further skills sheets, lists of objectives and practical work.

These are references to the Teacher's Resource Disk. The disk contains assessment criteria sheets, summary sheets, tests and foundation worksheets.

*CSF and NP*

For each unit the CSF and NP levels are clearly indicated, allowing you to link the text directly to the curriculum guidelines.

# Making a glossary

As you work through this book you will come across words in **bold type**. These are words that might be new to you, or common words that have a special meaning when used in science. They are also listed in the 'New words' section at the end of each unit.

To help you remember these words, you should make your own glossary of science terms. Set aside a section of your science workbook for your glossary. Leave half a page for each letter of the alphabet. As you come across new words, add them in alphabetical order to your glossary. Also include the meaning of the word, either by referring back to the text or by using a dictionary.

In this way you will build up a useful list of scientific terms.

## Using a dictionary

You probably already know how to look up words in a dictionary. For example, if you wanted to look up the meaning of the word 'silly', you would search for:

S
then Si
then Sil
and so on.

There are guide words at the top of each page to help you. For example, the page that contains 'silly' may have 'silent' at the top. Eventually you will find an entry something like the one shown opposite.

Silly has at least three meanings. Many scientific words have a **scientific use** as well as other uses. For example, a dictionary might give the definitions of the word 'cell' shown opposite.

Copy and complete the table below. Use a dictionary to find one scientific and one non-scientific use for each of the following words.

> **silly** *adjective*
> 1. showing a lack of good sense. *Usage:* 'the blow knocked me *silly*' (= stunned, dazed).
> 2. *Cricket:* (of a fielding position) very close to the batsman. *Word family:* **silly**, *noun*, a silly person. [Middle English *sely* happy]
> From *Heinemann Australian Dictionary* 3rd edition (Heinemann Education Australia, 1991).

> **cell**
> a very small room (non-scientific use)
> a compartment in a honeycomb (non-scientific use)
> a device for producing an electric current (**scientific** use)
> a microscopic unit of living matter (**scientific** use)
> a small group of people (non-scientific use)

| Word | Scientific use | Non-scientific use |
|---|---|---|
| control | | |
| element | | |
| symbol | | |
| reaction | | |
| formula | | |
| indicator | | |
| neutral | | |
| base | | |
| salt | | |
| tissue | | |

# CSF outcome statement

**Life and living**                                                           **Chapter**

LEVEL 3
- Map relationships between living things in a habitat. .................... 8

LEVEL 4
- Explain how animals use their senses to detect and respond to their environment. ............................................................................. 1
- Describe the functioning of the support, transport and reproductive systems in plants and animals and the respiratory systems in animals. ... 6
- Identify living and non-living things that affect the survival of organisms in an ecosystem. ..................................................................... 6, 8

LEVEL 5
- Explain how plants and animals obtain, transport and store nutrients. ... 7
- Recognise that all plants and animals are made up of cells and describe the major features of cells. ........................................................ 7
- Explain the various environmental changes on living things in ecosystems. ............................................................................. 8

LEVEL 6
- Describe the main features of DNA and its role in genetic continuity. .... 7
- Explain the role of living things in the cycling of matter and flow of energy in an ecosystem. ........................................................... 8

**Natural and processed materials**

LEVEL 4
- Identify factors that determine the choice of materials for particular purposes, including their suitability as fuels. .............................. 3, 5
- Distinguish between elements and compounds. .............................. 3, 4

LEVEL 5
- Give examples of common chemical reactions and describe them using word equations. ....................................................................... 4, 5
- Describe simple patterns in the arrangement of elements in the periodic table. ....................................................................................... 3
- Discuss the characteristics, including chemical reactions, of groups of similar substances. .................................................................... 3
- Describe how a change in reaction conditions influences the speed and product of the reaction. ............................................................. 4

LEVEL 6
- Discuss the characteristics, including chemical reactions, of groups of similar substances. .................................................................... 5

**The physical world**

LEVEL 3
- Identify a range of examples of energy transformations involving movement, light, sound and current. ........................................... 9

                                                                    **Chapter**

LEVEL 5
- Explain how machines are used to modify force and motion.   10
- Describe the relationship between the characteristics of sound, the vibration properties of the source, and the sound wave.   11

LEVEL 6
- Interpret the outcomes of a variety of simple investigations in terms of energy transformations and conservation of energy.   9

### Earth and beyond

LEVEL 3
- Discuss events caused by the tilt of the Earth's axis, including seasons and the length of daylight.   12

LEVEL 5
- Discuss events caused by the relative movement of the Sun, Moon and Earth.   12
- Explain the methods of formation and the uses of common types of rocks.   12

LEVEL 6
- Describe the use of satellites and space probes.   12

# National Profile outcome statement

**Life and living**      **Chapter**

LEVEL 3
- Identifies external and internal features of living things that work together to form systems with a particular function. ............. 1
- Maps relationships between living things in a habitat. ............. 8

LEVEL 4
- Explains the functioning of systems within living things. ............. 1, 6, 7
- Identifies events that affect balance in an ecosystem. ............. 8

LEVEL 5
- Presents evidence that plants and animals are made up of functional units called cells. ............. 7
- Describes the role of living things in cycling energy and matter. ............. 8

LEVEL 6
- Explains how living things obtain and transport nutrients, transform energy and manage wastes. ............. 6
- Describes how genetic continuity is maintained from generation to generation ............. 7
- Analyses the effects of environmental change on living things and ecosystems. ............. 8

**Natural and processed materials**

LEVEL 3
- Makes connections between the structure of common materials and their properties. ............. 3, 5, 13

LEVEL 4
- Uses models of the substructure of materials to explain their properties and behaviour. ............. 3, 4, 5, 13
- Recognises and describes conditions that influence reactions and change in materials. ............. 4, 13

LEVEL 5
- Assesses the effectiveness of materials used for particular purposes. ............. 3
- Uses simple models of atoms to explain chemical reactions. ............. 4

**Working scientifically**

LEVEL 4
- Identifies the information needed to make decisions about an application of science. ............. 1, 7
- Identifies factors to be considered in investigations, controls which may be needed and ways of achieving control. ............. 2, 4, 9
- Collects and records information as accurately as equipment permits and investigation purposes require. ............. 2, 3, 5, 9
- Draws conclusions linked to the information gathered and the purposes of the investigation. ............. 2, 9
- Reviews the extent to which conclusions are reasonable answers to the questions asked. ............. 2
- Identifies the information needed to make decisions about an application of science. ............. 2, 13

**Chapter**

LEVEL 5
- Identifies factors that influence people's perceptions of science. .................................................. 1, 2, 4, 6, 7, 12
- Selects an appropriate pathway for an investigation, given its purposes and the resources available. .............................. 2, 5, 9
- Uses instruments and techniques to provide accurate and reliable results. ................................................................ 2, 5
- Selects ways to present information that clarify patterns and assist in making generalisations. ................................. 3
- Proposes and compares options when making decisions or taking action. ............................................................... 12

LEVEL 6
- Uses information as a stimulus for further investigation or analysis. ............................................. 1, 2, 4, 5, 6, 7, 13
- Plans procedures to investigate hypotheses and predictions for situations involving few variables. ..................... 2, 5, 9
- Selects instruments and techniques to collect useful quantitative and qualitative information. ............................... 2
- Analyses costs and benefits of alternative scientific choices about a community problem. ................................... 9, 13
- Assesses conclusions linked to the information gathered and the purposes of the investigation. ............................. 9

## Energy and change

LEVEL 3
- Identifies the chain of sources and receivers of energy within systems. ........................................................... 9, 11
- Designs and describes ways of impeding the transfer of energy. ............... 11

LEVEL 4
- Identifies forms and transformations of energy in sequences of interactions. ................................................... 9, 10, 11
- Identifies processes of energy transfer and conditions that affect them. ...................................................... 9, 10, 11

LEVEL 5
- Explains energy input-output devices using concepts of work, force and power. ........................................................... 10
- Analyses energy transfers where the purpose is to apply a suitable force to achieve an outcome. ............................ 10

## Earth and beyond

LEVEL 3
- Illustrates patterns of change observable from the Earth caused by the relationship between the Sun, Earth and Moon. ............ 12

LEVEL 4
- Examines ways scientists investigate the Earth, the solar system and the universe. .................................................... 13

# Working scientifically

## section one
# Observing, recording and hypotheses

**1** Let's experiment

**2** Sun sense

chapter 1
# Let's experiment

## Introduction

**Outcomes**

At the end of this chapter you should be able to:
- Plan an experiment to investigate a hypothesis.
- Make predictions based on the information available.
- State the difference between a control and a variable.
- Examine an experiment and identify the controls and variables.
- Design an experiment that includes a control and some variables.
- Draw conclusions from information presented and/or researched.
- Draw conclusions in light of the aim of an experiment.
- Write up an experiment in the correct scientific format.

Your teacher may give you a copy of these Outcomes for your workbook.

### The senses

When scientists observe the world they use their senses. Which senses do scientists use? From your own experiences in science, you probably know that scientists use the senses of sight, hearing, touch and smell when it is not dangerous to do so. They rarely taste anything (unless they are engaged in food science). Some substances used in science would taste terrible; others are highly poisonous.

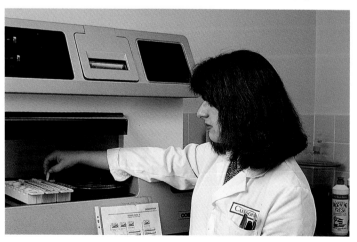

Once scientists have observed the world around them, they ask questions about what they have observed. They try to find answers to their questions by carrying out experiments or tests. This is called the **scientific method**. You will explore this method further in this chapter.

# Hypothesis-making and scientific method 1.1

## What is a hypothesis?

A **hypothesis** is a positive statement that can be tested. The statement is based on observations. A hypothesis is therefore an 'educated guess'. The scientific method is used to test a hypothesis.

These exercises will help you identify the steps in the scientific method.

1. The following statements are out of order. Copy them into your workbook in the order you think they should be.
   - Design and carry out the experiment to see if the hypothesis is supported.
   - Notice something – observe something happening, smell a strange smell, hear a strange noise.
   - Make conclusions, leading to further **predictions** and further tests.
   - Ask questions. Why did that sound occur? Why did that happen? What caused it?
   - From the statement, make predictions about what you think will happen. 'If my hypothesis is correct, then … will happen.'
   - If the hypothesis is not supported, review the hypothesis, change it, make new predictions and design new tests.
   - Form a hypothesis – a statement that tries to explain the observations made.

2. In your own words, summarise the steps in the scientific method.
3. Draw a flow chart to show how the steps in the scientific method are linked together.

## Controls and variables

When scientists design experiments they think very carefully about what they are going to do. In any experiment there are lots of things that can affect the results. These are called the **variables**. A scientist takes great care to test only one variable and **control** all the other variables.

For example, if you were going to test whether one athlete was fitter than another, you would have to put the athletes through exactly the same tests, under exactly the same conditions. If you didn't, your test would not be fair. You would have to make sure that all the things that could affect the results, other than the athletes' fitness, were kept **constant** (the same).

What's not fair about this experiment?

In many experiments, scientists must be able to compare what they are testing with something that doesn't change – a **control**. If you designed a test to see whether eating oranges stopped people getting colds, you might set up two groups. In one group, the people might eat five oranges a day. In the other group – the control group – the people would eat their normal diet. You would make a conclusion by comparing how many people from each group got colds.

**new words**

scientific method • hypothesis • prediction • variable • control • constant

# Spot the variables   1.2

1. Spot the variables in the photograph below. Look for variables that stay the same and variables that change.

   Copy and complete the table below for another three variables.

   | Variable | Stays the same | Changes |
   |---|---|---|
   | colour of coin | | |

2. Identify five variables in the photograph below. List them in a table like the one above.

3. Identify five variables in the illustrations below. List them in a table.

   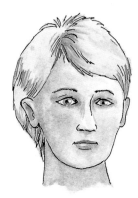 

**new words**

Add any new words to your glossary.

# Fair experiments 1.3

| NP | CSF |
|---|---|
| 4 | |
| 5 | |
| 6 | |
| 7 | |

Are these experiments fair?

1. Find as many variables as you can in these two examples. Record your answers in a table with headings: 'Variable', 'Stays the same', and 'Changes'.

   a

   b

2. These two girls decide to have a race to the nearest tree. It's not going to be a fair race. What variables should be controlled to make it fair?

Your teacher may give you more work on fair tests.

**new words**

Add any new words to your glossary.

# Recording the experiment

1.4

Scientists must record their results **accurately**. This enables them to **repeat** their results. It is only when an experiment can be repeated to produce the same results that valid conclusions can be made. It is for this reason that scientists write up their findings in a standard format.

1. Read the notes of a student's experiment below. Write down the student's aim, apparatus, method, results, discussion and conclusion.

○ My glass bottle of soft drink explodes when I put it in the freezer with the top on. Why does this happen? Why doesn't it happen when I put the bottle in the fridge?

○ Water expands when it is frozen and this causes the bottle to break.

○ If I place two bottles in the freezer, one with a lid and one without, the bottle with the lid will break because the water will expand as it freezes.

○ Take two glass soft drink bottles. Fill both bottles with water. Leave the lid off one bottle and put the lid on the other bottle. Put both bottles in the freezer.

○ The following things will make a difference to this experiment:
   • the amount of water in the bottle
   • the time in the freezer
   • the temperature of the freezer
   • the type of water
   • the size of the bottle.

○ I left the lid off one bottle to compare it with the bottle with the lid on.

○ The bottle with the lid on did explode. The ice in the bottle without the lid had pushed out the top of the bottle. My hypothesis was supported.

○ I could investigate whether all liquids behave in this way.

○ **Danger:** Do not try this yourself. It could be very dangerous.

2. State the variables that were controlled in this experiment.
3. State the one variable that was tested.
4. State what the control was, and explain why it was needed.
5. Describe what further experiments could be carried out.
6. List the student's initial observations and questions.
7. State what a standard practical report must contain.

# 1.4 continued...

**8** Another student made the following observation:

○ I have a pot of daffodils growing
○ inside, near the kitchen window. The
 daffodils closer to the window are
○ much bigger than the ones further
 from the window.
○

She then asked these questions:

○ Is this because some daffodils are
○ stronger than others?
 Or because they are getting more
○ light?
 Or because they are warmer?
○

From this information, write down: a hypothesis, a possible prediction, a method, a list of variables to be controlled, one variable to be tested, and any possible further investigations.

Your teacher may give you a handout that you can compare with your own work.

You will notice that scientists say that a hypothesis is **supported** rather than **proved**. As you know, scientific theories change over time, so it isn't possible to prove a hypothesis absolutely. With advances in technology, many things we thought were facts have been shown to be incorrect.

If there is enough evidence to support a hypothesis, it may become a theory, but it can still be changed if new evidence is found.

**new words**

accurately • repeat • support • prove

# Let's experiment
## 1.5

Your task is to design and carry out your own experiment.
   You must clearly state:
- the hypothesis you are investigating
- the variable to be tested
- the variable(s) that must be controlled
- the results of the experiment
- your conclusion.

You should think about the best way to present your results. They could be in words (**qualitative data**) or in numbers (**quantitative data**).
   Try to suggest how your research could be extended.

Your teacher may give you an assessment criteria sheet for practical reports.

### Hypothesis ideas

It is best to choose your own ideas to investigate. Looking through this chapter may give you some ideas. If you are having trouble, some suggestions are listed below.
1. Flowers kept in water that is changed daily will last longer than those kept in water that is not changed.
2. Athletic people have a better sense of touch and balance than non-athletes.

3. People who don't smoke have a better sense of taste and smell than those who do smoke.
4. Music makes plants grow faster.
5. Saccharin is a sugar substitute that tastes no different to sugar.
6. If you eat carrots, you can see better in the dark.
7. Using toothpaste cleans your teeth better than brushing alone.

Your teacher may give you some skills sheets to help you improve your practical report writing.

**words**
Add any new words to your glossary.

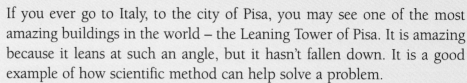

# The Leaning Tower of Pisa

**1.6**

| NP | CSF |
|---|---|
| 4 | |
| 5 | |
| 6 | |
| 7 | |

If you ever go to Italy, to the city of Pisa, you may see one of the most amazing buildings in the world – the Leaning Tower of Pisa. It is amazing because it leans at such an angle, but it hasn't fallen down. It is a good example of how scientific method can help solve a problem.

The tower was built at the end of the twelfth century, as part of the city's cathedral. It is about 55 metres high. Even when it was built, it had a slight lean to the south. When a bell tower was added in the thirteenth century it was slanted towards the north to try to correct the tilt of the tower. Even so, the tower has leaned a bit further every year, and the top is now about 5 metres away from vertical.

The main question scientists needed to answer was why the tower leant in the first place. They found that the tower rested on a layer of very soft clay 8–40 metres below the surface. The weight of the tower was compressing the clay on one side, causing the building to lean.

The next problem was to work out a way to stop the tower falling over. Steel cables wrapped around the tower were one possible solution. Another was to put heavy cement blocks on the north side of the tower to counter-balance the lean. However, both of these were only temporary solutions.

Scientists are still debating what to do to stop the tower falling over. Many ideas have been put forward. The first is to consolidate (compact) the soft clay to make it a harder base. This could be done by extracting water from the clay.

The rate of leaning seems to depend on the rate at which the city uses underground water. One suggestion is to build a dam wall to isolate the tower, so it would not be affected by pumping up the underground water.

### your turn

1. Describe the problem of the Leaning Tower of Pisa and its underlying cause.
2. Explain some of the solutions that have already been tried.
3. Describe another solution that has been proposed.
4. Apply the scientific method to the problem of the leaning tower. List the questions that were asked, the hypotheses that were made, and the methods used to solve the problem.
5. You might like to investigate another place in Italy that poses a problem for scientists – the city of Venice. Find out the problem facing Venice and how science is being used to overcome it.

### new words
**Add any new words to your glossary.**

# another look 1.7

## 1.1 Hypothesis-making and scientific method
1. In your own words, state what a hypothesis is.
2. Outline the scientific method. Briefly explain what each step should involve.
3. State what a variable is.
4. Describe what a control is, and explain why it is needed.

## 1.2 Spot the variables
These two plants were set up by a student to investigate the effect of light on the growth of plants.

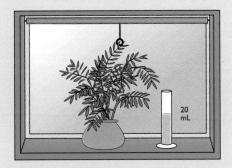

1. Explain whether this is a fair test.
2. List the variables involved in this test.
3. Explain how these variables can be controlled to make it a fair test.

## 1.3 Fair experiments
1. Explain why we must be able to trust the results of an experiment.
2. Explain how experiments should be designed so that they are fair.
3. Give an example of an experiment you performed, describing the variables you controlled and the variable you changed.

## 1.4 Recording the experiment
1. Explain why experiments must be able to be repeated.
2. List the set format that scientists follow.
3. Explain why scientists say that a hypothesis is supported, rather than proved.
4. Explain whether a theory can be changed.

## 1.5 Let's experiment
1. Briefly outline an experiment you designed and tested.

## 1.6 Science in Action: The Leaning Tower of Pisa

1. State why scientists are needed at this major tourist attraction.
2. Outline the problem scientists face and the possible solutions to the problem.
3. Using the leaning tower as an example, explain how science helps us in our daily lives.

### Good and bad learning behaviours

1. There are many ways to approach a task such as a project or revision for a test. In your workbook, draw up a table like the one below. List as many ideas as you can think of under each heading.

| Good things to do to complete a task | Bad things to do to complete a task |
| --- | --- |
| Make a time plan | Leave it to the last minute |
|  |  |

2. As a class, write all the ideas on the board. Discuss solutions to the problems in the 'bad things' column to help you complete your work in science.

### Strengths and weaknesses

At the end of a unit of study, it is important to be able to look back at what you have done to see whether you have completed all the tasks and understood everything. Try the following exercise.

1. Sit with a partner and go through your workbooks together. Make sure you have both completed all the homework set. Compare your books to see if there are any differences. Have you missed any work? Lost handout sheets? Do you need to catch up on any work? Were you absent and need to find out what was completed while you were away?

    Get your workbook up to date. Ask your teacher for any worksheets you are missing.

2. Read through your work. Ask yourself:
   - Have I understood everything that was covered?
   - What work didn't I understand? (Make a list of questions you would like to ask your teacher.)
   - What work did I complete well?
   - What work do I need to spend more time on?
   - Do I have everything I need in my workbook to study for a test?
   - Is there anything else I can do to bring my science workbook up to date?

   If you have any difficulties, see your teacher.

Your teacher may give you a summary sheet and a topic test.

# Skillbuilder: Making a bibliography

A **bibliography** is a list of all the books, magazines or other resources (such as CD-ROMs) that you have used to complete a written task. It is important to include a bibliography to show where your information and ideas have come from.

A bibliography always comes at the back of a piece of work and is always arranged in alphabetical order.

A bibliography must include the following information for each resource you have used:
- the name of the author or editor (surname followed by first name or initials)
- if it is a magazine or newspaper article, the title of the article (inside 'quotation marks')
- the full title of the book, magazine, newspaper or CD-ROM (underlined or in *italics*)
- if it is a book, the publisher of the book
- if it is a magazine, the volume and number of the magazine
- the date when the book, magazine, newspaper or CD-ROM was published.

Some examples are shown below.

| | |
|---|---|
| A book with one author | Garton, A. ScienceMoves 1. Rigby Heinemann, 1996. |
| A book with two authors | James, M. & Parsons, M. Heinemann Outcomes Science 1. Rigby Heinemann, 1995. |
| A book with no recorded author | The World Book Encyclopedia. 1981. |
| An article from a newspaper | Flynn, D. 'Big two battle to dominate Web.' The Age. 30 April 1996, p. D1. |
| An article from a magazine or journal | Low, Tim. 'Cures from the canopy.' Australian Natural History. Vol. 22, no. 4, Autumn 1987. |
| A CD-ROM | Compton's Interactive Encyclopedia. 1995 Edition. Compton's New Media, 1994. |

The author, title, publisher, date of publication and edition number, is usually found on the title page of the book or the next page. Always try to record this information as you are doing your research or taking notes.

1. Write a bibliography that contains an entry for four different kinds of resources.
2. Explain why it is important to record this information when taking notes.

## chapter 2
# Sun sense

## Introduction

> **Outcomes**
> At the end of this chapter you should be able to:
> - Describe the structure of the skin and label a diagram of the skin.
> - Describe the structure of the ear and label a diagram of the ear.
> - Describe the parts of the eye and label a diagram of the eye.
> - Explain long- and short-sightedness.
> - Find your blind spot.
> - Explain optical illusions.
> - Describe the structure of the tongue.
> - Carry out an experiment to show which areas of the tongue we use for taste.
> - Research the effect of the sun on our bodies.
> - Carry out a project on sun sense.
>
> Your teacher may give you a copy of these Outcomes for your workbook.

### How much do you know about the sun?

Do you go to the beach during the summer holidays? Do you lie in the sun to get a tan? Despite all the information we now have about the effects of the sun on our skin, many people still don't follow safe practices when they are outdoors.

In this chapter you will look at the sun's effects on your body by studying the skin and the senses. You will then examine why the senses are so important to scientists.

 **your turn**

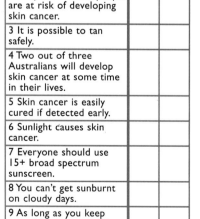

| Statement | True | False |
|---|---|---|
| 1 Skin cancers only occur in older people. | | |
| 2 Only fair-skinned people are at risk of developing skin cancer. | | |
| 3 It is possible to tan safely. | | |
| 4 Two out of three Australians will develop skin cancer at some time in their lives. | | |
| 5 Skin cancer is easily cured if detected early. | | |
| 6 Sunlight causes skin cancer. | | |
| 7 Everyone should use 15+ broad spectrum sunscreen. | | |
| 8 You can't get sunburnt on cloudy days. | | |
| 9 As long as you keep reapplying sunscreen, it will protect you all day. | | |
| 10 A suntan is healthy. | | |

Source: Anti-Cancer Council of Victoria.

1. Read the statements opposite. State whether you think they are true or false. You may like to ask members of your family for their opinions too.
2. Combine the answers of all class members using a tally sheet. As you work through this chapter, you may like to change some of your answers.
3. Now write down all the things you would like to know about the effects of the sun on the body.

# The skin         2.1

The skin is where the effects of the sun on our bodies are most noticeable. Let's investigate the skin.

Look at the diagrams of the skin below. Using these diagrams and what you already know about the skin, try to answer the following questions. If you have trouble with a question, discuss it with your teacher or carry out some research of your own.

Your teacher may give you a worksheet with diagrams of the skin.

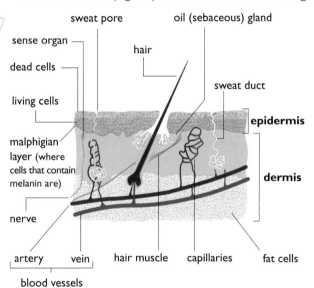
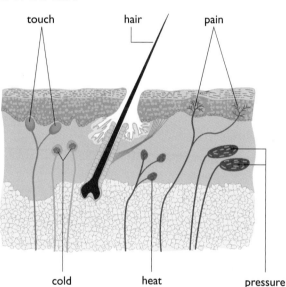

1  Stick a piece of sticky tape to the back of your hand and then pull it off.
   a  Describe the pattern on the tape.
   b  Explain why you didn't bleed when you pulled the tape off.
2  a  Explain why it doesn't matter if we lose the top layer of our skin. Is it alive?
   b  Name the layer of the skin that contains dead skin cells. (You will examine cells in more detail in chapter 7.)
3  Explain why it is an advantage to have nerves in the dermis.
4  Explain why blood vessels are found in the lower layers of the skin.
5  Describe how germs could get into our skin.
6  State where the receptors are in the skin.
7  Describe and explain what the sweat ducts do.
8  The blood vessels in the skin can **dilate** (get bigger) or **constrict** (get smaller).
   a  Explain how you think this would help to cool and heat the body.
   b  Explain what part the hairs on your body play in this.

# 2.1
*continued...*

### on your own

## Investigate sensitivity

Design an experiment to test one of the following hypotheses. (First you may want to review what a hypothesis is.)

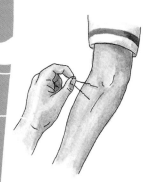

❶ We are more sensitive to touch on our hands than on any other part of our bodies. **Clue:** Use the two points of a hairpin. Try touching it on different parts of the body with the points spaced 20 mm apart, 10 mm apart and 5 mm apart. Record your results in a table like the one below. The skin is very sensitive if it can feel two separate points close together.

| Part of the body tested | Number of points you could feel | | |
|---|---|---|---|
| | Points 20 mm apart | Points 10 mm apart | Points 5 mm apart |
| forearm | | | |
| back of neck | | | |
| lip | | | |
| forehead | | | |
| fingertip | | | |

❷ The skin will react differently to heat and cold. **Clue:** Make a person's skin hot and then make it cold. Observe the effects.

**Remember:** To be valid, an experiment must be able to:
- be repeated
- withstand **criticism**
- be controlled.

Only one variable must be tested. If you have forgotten how to design an experiment, refer to chapter 1 before continuing.

## Investigate sunburn

There are three **degrees** of burns:
- **superficial** burns, which affect only the outer layers of the skin
- **intermediate** burns, which cause blisters
- **deep** burns, which affect all the layers of the skin.

❶ Find out which degrees of burns are commonly caused by sunburn.
❷ Find out how you would treat somebody who had sunburn.

### new words

dilate • constrict • criticism • degree • superficial • intermediate • deep

# It's great to see you

## 2.2

## Parts of the eye

1  Copy and complete this passage, referring to the diagrams of the eye below. Light enters the eye through the transparent conjunctiva and then travels through the watery part of the eye. This water bends the light through the _____ into the _____ . This directs the image onto the back of the eye. No light goes through the white of the eye – the _____ . Light hits the _____ at the back of the eye where there are receptor cells. They send messages via the _____ nerve to the brain. If light does not fall on the retina, a person will not be able to see correctly. Where the optic nerve leaves the back of the eye there are no receptor cells. This is called the _____ _____ . If light falls here, _____ image is formed.

Your teacher may give you a scrambled eye puzzle.

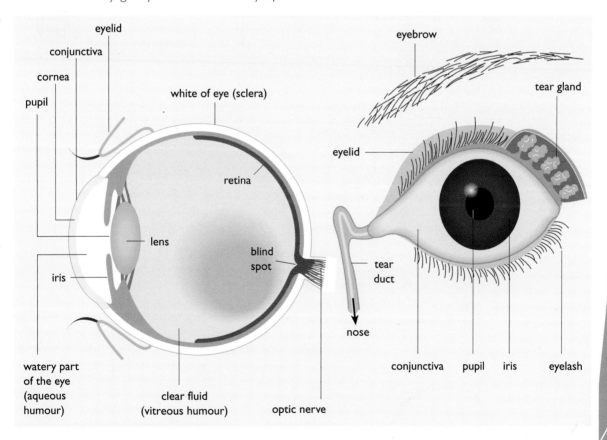

2  Name:  **a** the coloured part of the eye  **b** the hole in the eye.
3  State what stops sweat from your forehead running into your eye.
4  Describe the function of the eyelashes.
5  Explain when and why tears are produced.

Sun sense  **17**

## 2.2 continued...

### Seeing is believing

1. Look at the diagrams below. In your own words, describe how the lens and eye change when we look at near and distant objects.
2. Explain why it is important for the lens to change shape. **Clue:** Think about the image formed on the retina.

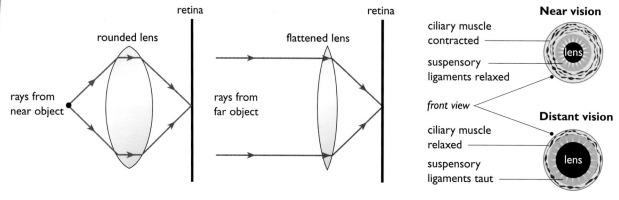

3. Look at the diagrams which show the effects of short- and long-sightedness.

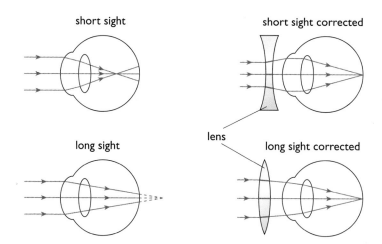

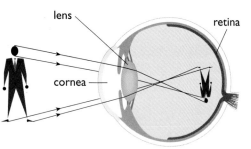

a Explain why long- and short-sighted people see a blurred image.
b Describe the difference between long- and short-sightedness.
c Describe how short-sightedness is corrected.
d Describe how long-sightedness is corrected.

When light passes through the eye, the image that is formed on the retina is not what you would expect. Look at the picture opposite. State two ways in which the image formed on the retina is different to the real image.

18  ScienceMoves 2

### try this

### Let's look inside

Your teacher will demonstrate how to dissect a bull's eye. The following steps will help you to dissect a bull's eye yourself.

① Locate the optic nerve at the back of the eye.
② Hold the eye firmly. With the scalpel, cut around the optic nerve in a wide arc.
③ Gently press the eye so that the fluid is squeezed into the petri dish.
④ Examine the fluid to find the lens – a hard, clear piece from the centre of the fluid.
⑤ Now try to turn the eye inside out. Examine the cornea and the coloured cells of the retina. (Ask your teacher if you have trouble finding these.)

1. Describe the parts of the eye you found.
2. Name the fluid that came out of the eye.
3. Explain your feelings about the dissection. Was it worthwhile?

**COLLECT**
- bull's eye (fresh not frozen)
- sharp pointed scissors
- razor blade
- tweezers and probe
- newspaper
- lab coat
- gloves
- petri dish

### on your own

### Looking into eyes

Prepare a report to present to the class on one of the following topics.
① Investigate optical illusions and the blind spot.
Your teacher may give you some material on this topic.
② Design an experiment to see what effects light has on the pupil. The picture opposite and table below may give you some clues as to how to carry out this experiment.
③ Using an eye chart, test class members to see who has the best eyesight. Describe how you can make sure this is a fair test.
④ Investigate the different brands of sunglasses available to see which blocks out most UV light.
⑤ Investigate diseases of the eye such as cataracts, which cause the cornea to become cloudy, and conjunctivitis (inflammation of the conjunctiva).

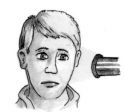

| Light condition | Size of right pupil | Size of left pupil |
|---|---|---|
| Dim light | | |
| Torch light | | |
| Just after torch light | | |
| Just after blindfold removed | | |

### new words

Add all the new words describing the eye to your glossary.

# It's a matter of taste 2.3

| NP | CSF |
|----|-----|
| 3  |     |
| 4  |     |
| 5  |     |
| 6  |     |

Your senses of taste and smell depend on special cells called **receptors** in your mouth and nose. The receptors in the mouth are called **taste buds**. They are found on the tongue and on the roof and floor of the mouth and throat. The taste buds send information through the **cranial nerve** to the brain. In the nose, the smell receptors are found on the **olfactory membrane**, which is linked to the brain by a nerve.

Humans can detect more than 3000 different chemicals by their smell.

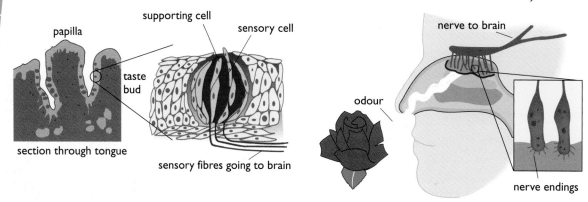

### try this

**COLLECT**
- blindfold
- sweet solution
- salty solution
- sour solution (lemon juice or vinegar)
- bitter solution
- distilled water (control)

## Sweet and sour

The taste receptors on the tongue can recognise four main tastes: sweet, sour (acid), salty and bitter. Different parts of the tongue are sensitive to each of these tastes. The chemicals in foods first have to dissolve in the **saliva** on the tongue before they can be tasted.

❶ Design an experiment to test whether the 'map' (right) of the tongue's taste receptors is correct.

❷ Carry out your experiment. Sample as many class members as possible. Make sure you use clean equipment and new solutions for each person.

❸ Design an experiment to test whether we use our sense of smell to help us taste food. Record your results in a table like the one shown below.

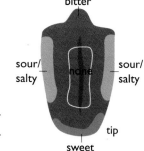

| Sample | Guess (with sense of smell) | Guess (without sense of smell) | Actual food identity |
|--------|-----------------------------|--------------------------------|----------------------|
| Food A |                             |                                |                      |
| Food B |                             |                                |                      |
| Food C |                             |                                |                      |

### new words

Add any new words to your glossary.

# project
# I hear you

## 2.4

Your task is to find out about the different parts of the ear, and explain how we hear and balance. The diagrams and questions below will help to get you started.

Your teacher may give you an enlarged copy of these diagrams.

| NP | CSF |
|----|-----|
|    | 3   |
|    | 4   |
|    | 5   |
|    | 6   |

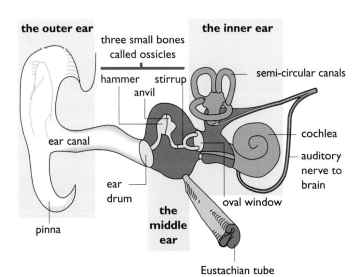

1. Name the part of the ear that collects sound and channels it inside the ear.
2. Describe the structure and function of the ear drum. State what happens if the ear drum is damaged.
3. Name the three small bones and describe how they help us to hear.
4. Describe how sound information is sent to the brain. Name the nerve involved.
5. Explain how the ear helps us to balance.
6. Describe the difference between the inner, middle and outer ear.
7. Describe the Eustachian tube and why we need it. How is it involved in the sensation we feel when our ears 'pop'?

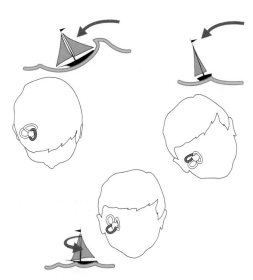

Movement is detected by different parts of the semi-circular canals. The shaded area shows where the head movement is detected.

Your teacher may give you an assessment criteria sheet and ask you to provide a bibliography. See the Skillbuilder at the end of chapter 1 if you need help in preparing the bibliography.

### new words

**Add any new words to your glossary.**

Sun sense **21**

# Can you tan safely?

## 2.5

| NP | CSF |
|---|---|
| 4 | |
| 5 | |
| 6 | |
| 7 | |

**Skin cancer** is a disease that can be **fatal**. There are different types of skin cancer. They appear on the skin in the form of a mole or mark that changes colour or shape. Anyone can develop skin cancer, but if it is detected early enough it can be cured.

The amount of **exposure** a person has to the sun during their childhood is an important factor in whether they will develop skin cancer at a later age. People can develop skin cancer from **adolescence** onwards, no matter what type of skin they may have.

Tanning is your body's **defence mechanism** against the sun. The sunlight triggers the body to produce more **melanin** (a brown pigment) to protect the skin from further damage. Exposure to the sun also causes your skin to become thicker and wrinkles to develop faster. Unfortunately, because of the number of fair-skinned people in Australia and our outdoor lifestyle, two out of every three Australians will develop skin cancer.

The damaging part of the sun's light is the **UV** (ultraviolet) radiation. UV radiation is invisible, and is present even on cloudy days. The **ozone layer** is a layer of gas in the Earth's atmosphere that helps to screen out damaging UV radiation. Because of damage caused to the ozone layer by natural and industrial chemicals, more UV radiation is reaching us today than it did 20 years ago. This means that skin cancers are likely to become even more common, unless we change our behaviour.

Science has certainly had an influence on our behaviour. Advertising images of 'healthy' tanned bodies are being replaced by education campaigns that teach us to be more 'sun smart'. Scientists have also developed sunscreens – creams that can effectively block out ultraviolet rays. Everyone should wear a hat and a shirt and apply a 15+ sunscreen whenever they are outside in the sun. Sunscreens should be reapplied every two hours because they only protect for a limited period.

**Melanoma:** the rarest and most dangerous skin cancer.

Appearance: a new spot, a freckle or a mole that changes colour, thickness or shape over a period of months. Often dark brown to black, red or blue-black. Can appear anywhere on the body.

Affects all ages but is rare in children.

Early treatment results in total cure in 95% of cases.

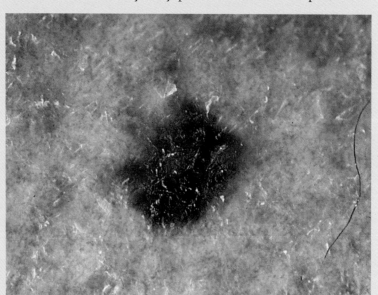

1. Look back at the questions you answered at the beginning of this chapter. Did you get them all right? Change any of your answers if necessary.
2. Describe what you can do at your school to change people's ideas about tanning.
3. Describe what your school is doing to protect students from sun exposure.

## on your own

### Your skin and the sun

Investigate one of the following and prepare a short report.

1. Burning, freckles and tanning: Investigate what happens to the skin and why tanning occurs.
2. The composition of light: Investigate sunlight, ultraviolet light, the light spectrum, and forms of radiation.
3. Skin cancer: Describe skin cancer, what happens to the skin and what the treatment is.

**Squamous cell carcinoma (SCC):** Moderately dangerous, represents 20% of all skin cancers. Most common in people aged over 40.
Appearance: scaling, red areas which may bleed easily or ulcerate, with appearance of open sores.
Grows rapidly over a few months.
Appears on sites most often exposed to the sun.

4. The ozone layer: Investigate what the ozone layer is, what is happening to it, and how this will affect us in the future.
5. Sunscreens: Investigate what chemicals are used in sunscreens and how they work. Design an experiment to test which sunscreen is the best.

## new words

skin cancer • fatal • exposure • adolescence • defence mechanism • melanin • UV • ozone layer • melanoma • squamous cell carcinoma

# another look 2.6

## 2.1 The skin

1. Describe the different parts of the skin, and what you know about each part.
2. Describe the different functions of the skin.
3. Which would we feel first – pain or heat? Why? Explain why you think the body works like this.
4. Copy and label the following diagram.

## 2.2 It's great to see you

1. Copy and label the following diagrams.

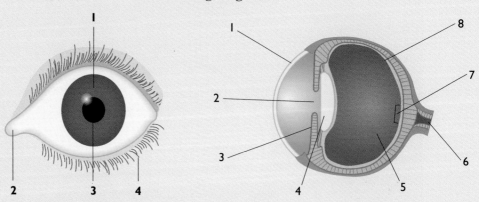

2. Explain how light travels through the eye to the retina.
3. Explain why we have a blind spot.
4. Describe how the image that forms on the retina is different from the real object.
5. Describe the difference between long- and short-sightedness.
6. Describe how long- and short-sightedness can be corrected.
7. Describe the dissection of a bull's eye and what the parts of an eye actually look like.

## 2.3 It's a matter of taste

1. Explain how we taste and smell.
2. Describe how the taste buds are arranged on the tongue.
3. Explain how our sense of taste is influenced by our sense of smell.

# another look

## 2.4 I hear you

1 Copy and label the diagram of the ear.
2 Briefly explain how sound travels through the ear.
3 Describe how the ear is responsible for balance.
4 Explain why our ears 'pop' when we go up a steep hill or fly in an aeroplane.

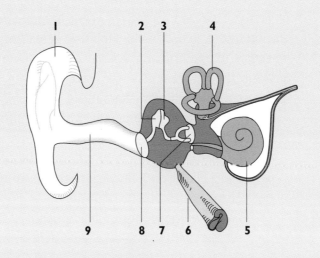

## 2.5 Science in Action: Can you tan safely?

1 Write a letter to a friend who loves to lie in the sun at the beach during the summer. Try to convince him or her about the dangers of tanning.

## Word find

See how many words you can find in the grid below. When you have finished, you could make your own word find and swap it with a friend.

```
A S T I G M E B U T N A I H C A T S U E
S E M I C I R C U L A R C A N A L S G V
E Y E S T R A I N O T S I G O L O T O R
V P H C E E P S M S I T A M G I T S A E
R S B I F O C A L C O P T I C I A N L N
E S S U M S I B A R T S S E S S A L G C
N E O D A O P T N O I T C E F N I O I I
Y N R E M M A H O P L S O U N D B R G T
R D E A O R M Y W A X I Y E P L I P U P
O N E F A N O S N P U R R I T S O L R O
T I S C R I C A N A L T G A E L H C O C
I L T P A A U H T G O E A R D R U M E A
D B E Y E R A I P O Y M C E G A M I N N
U O T N Y B L N E S I O N O P T I V C I
A P R O S P G L I G H T H E A R I N G T
N O I S I V S T I E S P O C U L I S T E
C H Y P E R O P I A N O I T C A R F E R
```

**WORDS TO FIND**
anvil
auditory nerve
blindness
brain
canal
cataract
cochlea
cornea
deaf
ear
eardrum
Eustachian tube
eye
glasses
hammer
hearing
image
infection
iris
lens
light
noise
optician
optic nerve
pupil
retina
see
semicircular canals
sound
speech
sties
stirrup
vision
wax

Your teacher may give you a summary sheet and a topic test.

# Skillbuilder
## Reviewing your work

When you submit a piece of work it should not be full of mistakes. It is important to get into the habit of rereading your work to correct any mistakes in spelling, expression and punctuation. See how well you can pick up the mistakes in these three exercises.

1. Correct the following sentences.
   - The lends locks like jelly with little blood vessels the pupil is a big black hole.
   - The lens bend the light entering the eye.
   - I was glad I done it. It gave me lot of courage to dissect the eye.
   - The eye look like it was starring at you.
   - The optic neve looked like a peace of fat.
   - We are going to do an experiment of trying to investigate what the eye woks like.
   - The lens were a clear roung blob.

2. Correct the following spellings:

   | | | | |
   |---|---|---|---|
   | rectengular | oposite | traveled | lends |
   | vains | coulours | conclution | differon |
   | cojunctiver | dicection | puple | functon |
   | diagnally | splites | rentina | midle |
   | realy | pencile | reflexed | dysection |
   | somtimes | parallelle | beacker | positition |
   | flud | exprience | discussting | peice |

3. Rewrite the following sentences, correcting all the mistakes.
   - The eye looked like a round object with a big piple which was blue, the conjectiva was all like shrivelled up, the colour of the optic nerve was orangery.
   - The eye looked like a normal eye the conjectiva look like a clear cover over it.
   - The corna was round and clear and you could see your finger throughout it.
   - The lens was clear and the pupol and the iris you could see because it was black.

Natural and processed materials

## section two
# Elements and compounds

*science in context*
**3 Elements around us**

**4 Elements make compounds**

**5 Acids and bases**

# Chapter 3

# Elements around us

## Introduction

**Outcomes**

At the end of this unit you should be able to:
- Describe an element in terms of particles and atoms.
- State some common elements and their uses.
- Identify some elements and devise a key for the elements.
- State the difference between a metal and a non-metal.
- State some of the properties of elements.
- Describe metals, giving examples.
- State where you would find some common metals.

Your teacher may give you a copy of these Outcomes for your workbook.

This section is all about different substances that we use every day. Scientists divide these substances into two groups: elements and compounds.

**Elements** are substances that are made up of only one type of particle, or **atom**. Substances such as hydrogen, oxygen and calcium are elements. When different elements join together they form **compounds**. Elements and compounds are around us everywhere.

In chapters 3 and 4 you will investigate elements and compounds. In chapter 5 you will look at two common types of compounds, called acids and bases.

Making margarine.

### Hydrogen

Hydrogen is an element: it contains only hydrogen atoms. Hydrogen has many uses.

Filling weather balloons.

# Meet the elements

## 3.1

**on your own**

### It's elementary

Around the room your teacher has placed the following elements: sulfur, aluminium, carbon, iodine (crystals in a stoppered flask), silicon, tin, lead, gold, magnesium, calcium, iron.

1. Look carefully at each element and describe its appearance in terms of:
   - colour
   - liquid, solid or gas
   - shiny (lustrous) or dull
   - crystals, powder, small grains or large grains.

2. Find out the boiling point and melting point of each element on display.
3. Find out the main uses of the elements on display. Put your results in a table like the one below.

| Element | Appearance | Melting point (°C) | Boiling point (°C) | Main uses |
|---|---|---|---|---|
| Chlorine | Yellow-green gas with an unpleasant acrid smell. | −101 | −35 | Used in making many chemicals, including insecticides, synthetic rubber and plastics, and for bleaching and disinfecting. |
| Oxygen | Colourless, odourless gas. | −219 | −183 | Makes up 21% of the air. Essential to life. Used in respiration and burning, as fuel in rockets and welding. Used in hospitals, and in making steel. |

4. Design a key to classify the elements on display. (If necessary, review making dichotomous keys in *ScienceMoves 1*.)

**new words**

element • atom • compound

Elements around us **29**

# Where do you find the elements? 3.2

| NP | CSF |
|----|-----|
| 3  |     |
| 4  |     |
| 5  |     |
| 6  |     |

Elements are everywhere; we come across them every day. To make it easier to identify and use the elements, scientists have arranged them in a table called the **periodic table**. Each element is made up of only one type of atom. There are 92 naturally occurring elements and some others that have been made by scientists.

## The periodic table

1. In the periodic table, each element is represented by a **symbol**. Why do you think these symbols are used?
2. Copy the following list, then underline the elements.

    water          vinegar        bread
    sugar          oxygen         zinc
    air            brass          silver
    salt           neon           carbon

1. **a** List the names and symbols of five elements that you recognise from the periodic table.
   **b** Write down three facts about each one.
2. State where the metals are found in the periodic table.
3. Use the symbols of the elements to crack this coded message. Write the message in English.
   scandium + iodine + (europium − uranium) + nitrogen + cerium
   iodine + sulfur
   fluorine + uranium + nitrogen
4. Now write your own short message in 'element code'.

|  |  |  |  |  |  |  |  | 2 He<br>helium |
|---|---|---|---|---|---|---|---|---|
|  |  | 5 B<br>boron | 6 C<br>carbon | 7 N<br>nitrogen | 8 O<br>oxygen | 9 F<br>fluorine | 10 Ne<br>neon |
|  |  | 13 Al<br>aluminium | 14 Si<br>silicon | 15 P<br>phosphorus | 16 S<br>sulfur | 17 Cl<br>chlorine | 18 Ar<br>argon |
| 28 Ni<br>nickel | 29 Cu<br>copper | 30 Zn<br>zinc | 31 Ga<br>gallium | 32 Ge<br>germanium | 33 As<br>arsenic | 34 Se<br>selenium | 35 Br<br>bromine | 36 Kr<br>krypton |
| 46 Pd<br>palladium | 47 Ag<br>silver | 48 Cd<br>cadmium | 49 In<br>indium | 50 Sn<br>tin | 51 Sb<br>antimony | 52 Te<br>tellurium | 53 I<br>iodine | 54 Xe<br>xenon |
| 78 Pt<br>platinum | 79 Au<br>gold | 80 Hg<br>mercury | 81 Tl<br>thallium | 82 Pb<br>lead | 83 Bi<br>bismuth | 84 Po<br>polonium | 85 At<br>astatine | 86 Rn<br>radon |

| 63 Eu<br>europium | 64 Gd<br>gadolinium | 65 Tb<br>terbium | 66 Dy<br>dysprosium | 67 Ho<br>holmium | 68 Er<br>erbium | 69 Tm<br>thulium | 70 Yb<br>ytterbium | 71 Lu<br>lutetium |
|---|---|---|---|---|---|---|---|---|
| 95 Am<br>americium | 96 Cm<br>curium | 97 Bk<br>berkelium | 98 Cf<br>californium | 99 Es<br>einsteinium | 100 Fm<br>fermium | 101 Md<br>mendelevium | 102 No<br>nobelium | 103 Lr<br>lawrencium |

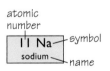

atomic number — symbol — name (11 Na sodium)

# Elements inside us

There are many elements in the human body. Iron is needed for red blood cells, oxygen is needed for respiration, and the carbohydrates we eat are made up of carbon, hydrogen and oxygen.

Your task is to prepare a poster showing elements that can be found in the human body. Show as many elements as you can find and state what each element is needed for.

Your teacher may give you an assessment criteria sheet or ask you to prepare your own assessment criteria using the procedure outlined on page 66.

**new words**
**periodic table • symbol**

# Introducing the metals     3.3

As you can see from the periodic table, metals are elements – they are made up of only one type of atom. Elements such as metals can be grouped according to:
- their **appearance** – what the metal looks like
- their **properties** – what the metal does
- their **uses** – what the metal is used for.

**METAL X**
**Appearance**
- red brown solid
- shiny and smooth
- sharp edges

**Properties**
- melts at 1077°C
- boils at 2567°C
- good conductor of electricity and heat
- soft (can be cut by scissors)
- 1 $cm^3$ weighs 8.9 g
- blackens when heated in air
- nothing happens when put into hydrochloric acid

**Uses**
- electrical wiring, jewellery, coins, water pipes
- mixed with other metals to make alloys like bronze and brass

Can you identify metal X from the information given?

### try this

**COLLECT**
- safety glasses
- strip of iron wire
- strip of magnesium ribbon
- sandpaper
- arm balance or electronic balance
- battery
- light globe and holder
- connecting wires

### Investigating metals (1)

Carry out the following investigations, first with iron and then with magnesium. Write down your results.

1. Describe its appearance.
2. Measure its mass.
3. Find out its melting point.
4. Test to see whether it is a good conductor of heat.
5. Test to see whether it is a good conductor of electricity.
6. Test to see how easily it breaks.

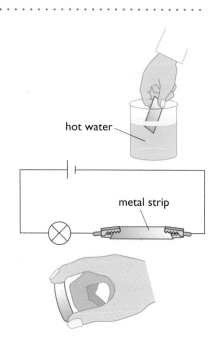

### try this

## Investigating metals (2)

Use one strip of each metal for each of the following experiments.

**COLLECT**
- strip of iron wire
- strip of magnesium ribbon
- dilute hydrochloric acid (2M HCl)
- test tube
- Bunsen burner
- heat-proof mat
- tongs
- safety glasses

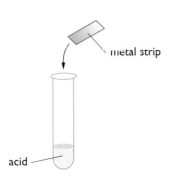

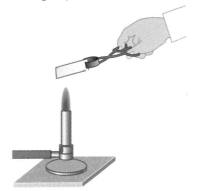

❶ Describe the effect of acid on the metal strip.

❷ Describe the effect of heat on the metal strip. Heat the metal strongly for about a minute.
**Caution:** Do not look directly at the flame.

1 Read the description of metal X again.
  Write similar descriptions for iron and magnesium using the same headings.
2 From your results, write down one possible use for each metal.

## Investigating metals (3)

❶ Find out the properties of the metals zinc and copper. Use the same tests that you used in the previous two experiments.

**COLLECT**
- sharp nail
- connecting wire
- battery
- light globe and holder
- 2M HCl
- strips of copper foil
- strips of zinc foil
- tongs
- safety glasses
- Bunsen burner
- heat-proof mat

❷ Write a description of each of these metals.

### new words

appearance • property • use

# How metals are used     3.4

| NP | CSF |
|----|-----|
| 3  |     |
| 4  |     |
| 5  |     |
| 6  |     |

Two properties of metals make them very useful. They are **ductile**, which means they can be pulled into fine strands to make wires. They are also **malleable**, which means they can be bent, shaped and made into sheets.

Some metals make stronger wires than others. Some metals change shape more easily than others.

## try this

**COLLECT**
- safety glasses
- 10 cm lengths of thin wire made from copper, nichrome, steel, magnesium, aluminium
- strips of different metals
- retort stand
- child's metal sand bucket
- 100 g masses
- sponge

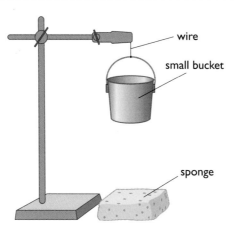

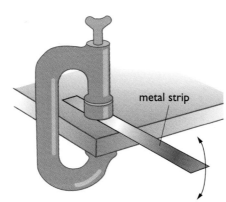

### Investigating metals (4)

**①** Design and carry out an experiment to compare the strength of the different wires. Remember to make sure your comparison is fair. The diagram opposite shows one possible method.

**②** Design and carry out an experiment to compare how well the strips keep their shape. Make sure that your comparison is fair. The diagram opposite shows one possible method.

For each experiment:
- draw your experiment
- present your results in a table
- describe how you made the comparison fair.

**new words**
ductile • malleable

34   ScienceMoves 2

# Metals and non-metals  3.5

Metals have some common properties.

| Metals | Non-metals |
|---|---|
| Metal elements are on the left-hand side of the periodic table. | The 21 non-metallic elements are on the right-hand side of the periodic table. |
| They are all solids at room temperature, except mercury. | Only nine non-metals are solids at room temperature. |

| NP | CSF |
|---|---|
| 4 | |
| 5 | |
| 6 | |
| 7 | |

Metals have other important properties that can be found by experiment.

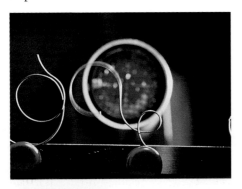

Some important properties of metals.

### try this

**Metal or non-metal?**

1. Divide the elements into two sets by appearance only.
2. Carry out experiments to find out:
   a  which elements conduct electricity
   b  which elements conduct heat easily
   c  which elements break into smaller pieces easily
   d  which elements change shape with heat or impact.
3. Record your results in a table.

1. Look at the periodic table. List the elements you tested under two headings: metals and non-metals.
2. Describe the following properties of the elements in the metal group: appearance, whether it conducts electricity and heat, strength, whether it changes shape with heat or impact.
3. Describe the same properties of the elements in the non-metallic group. (**Note:** There is at least one unexpected result in this group.)
4. Choose one property of each group. Suggest what an element with this property could be used for.

**COLLECT**
- samples of solid elements (copper, iron, magnesium, zinc, sulfur, carbon, iodine)
- battery or power pack
- connecting leads
- light globe and holder
- thermometer
- heat-sensitive paper
- test tube
- mortar and pestle
- electric kettle
- hammer

### new words
metal • non-metal

# The Earth's resources   3.6

| NP | CSF |
|----|-----|
| 3  |     |
| 4  |     |
| 5  |     |
| 6  |     |

Elements are found in many places, either on their own or joined with other elements to form compounds. Naturally occurring elements can be changed to make something new. The elements found in the Earth are called resources because they can be used to make useful things. The Earth is very rich in resources. Some examples are shown in the table below.

| Location | Resource | Raw material |
|----------|----------|--------------|
| Overground | Air | Oxygen, nitrogen |
| On the ground | Living things, water, soil | Wool, wood, fuel |
| Underground | Rocks | Metals, minerals, fossil fuels |

The story of how a resource is used is shown in this flow diagram.

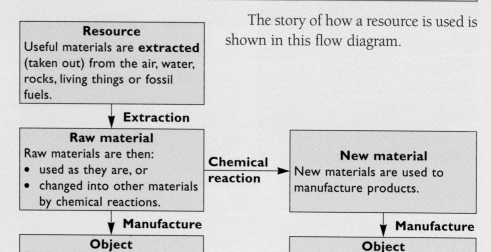

The story of iron can be shown in a similar flow diagram.

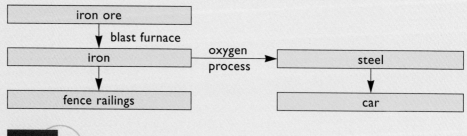

1. Give definitions of the following words: resource, extracted, manufacture.
2. State where useful materials can be extracted from.
3. Describe the story of iron in words.
4. Draw flow diagrams to show how the following things are manufactured:
   **a** a wooden fence **b** a gold ring **c** a clay plant pot.

**new words**

Add any new words to your glossary.

# another look 3.7

## 3.1 Meet the elements
1 Describe some common elements, their appearance and their uses.
2 Describe the difference between an element and a compound.
3 State another name for a particle.

## 3.2 Where do you find the elements?
1 Describe the periodic table.
2 List the first 10 elements in the periodic table and their symbols.
3 List 10 elements found in the human body and state what each element is needed for.

## 3.3 Introducing the metals
1 Describe the appearance and properties of copper, iron, magnesium and zinc.
2 Describe what tests you would carry out to demonstrate the properties of different metals.

## 3.4 How metals are used
1 Describe two properties of metals which make them useful.
2 Explain how you would design an experiment to show the useful properties of metals.

## 3.5 Metals and non-metals
1 Describe the difference between a metal and a non-metal.
2 List five metals and five non-metals.
3 State where you would find the metals in the periodic table.

## 3.6 Science in Action: The Earth's resources
1 Describe a resource. List some of the Earth's resources that we use.
2 Describe a raw material and explain how it is changed into something different.
3 Describe a chemical reaction.
4 Describe what is meant by the word 'manufacture'.
5 Put these words in the order in which the processes would normally occur: manufacture, resource, extracted, chemical reaction.

## 3.7 continued...

### Know your elements

Copy and complete these tables.

| Name | Symbol | Atomic number |
|---|---|---|
| Sodium | | |
| | Cu | |
| | | 6 |
| | Ag | |
| | | 17 |

| Substance | Appearance | Properties | Uses |
|---|---|---|---|
| Oxygen | | | |
| Hydrogen | | | |
| Iron | | | |
| Zinc | | | |
| Copper | | | |

### What's in a name?

1. Find the names of the elements that have been named after planets.
2. One element is named after an asteroid called Ceres. Name this element.
3. The following famous scientists have been honoured in the names of elements: Curie, Mendeleyev, Einstein, Nobel, Lawrence. Name the elements that carry their names and explain why these people are famous.
4. List the elements named after parts of the world.

### Element puzzles

1. Design your own 'Element Bingo' using the periodic table and the elements. One person calls out the name of an element and the others must cover that name with the correct symbol.
2. Think up your own element quiz questions. Here are a few examples.
   - Name the element used in light bulbs.
   - Name the element used in fluorescent lights.
   - Name the element used as a disinfectant in swimming pools.
   - Name the only metal that is a liquid at room temperature.
3. Design your own element crossword puzzle.
4. Design your own element word find puzzle.

Your teacher may give you a word find puzzle to do, and a summary sheet and topic test.

# Writing skills: using fewer words in sentences

Good notes use as few words as possible. For example, the following sentence is very long. It contains words that are not needed in a **scientific** note.

Some of the <u>acid was added to</u> the <u>powdered chalk</u> while I was standing at the bench <u>and the</u> whole <u>mixture fizzed</u> up and bubbles and froth showed.

The important words are underlined. The sentence could be rewritten as:

Acid was added to the powdered chalk and the mixture fizzed.

An even shorter version would be:

Acid + powdered chalk gives fizz.

Rewrite the following sentences using as few words as possible. Try to keep the scientific facts and leave out the rest.

1. We watched the lion while we were eating our ice-creams and it spent most of the time resting, almost as if it was watching TV.
2. Fine sand is used to dry flowers because it absorbs moisture and it can be collected when the tide in the bay is out.
3. To make the cement that you buy in cement bags at hardware stores, you need to mix powdered limestone and shale and heat the mixture in a kiln to 800°C, which is as hot as a Bunsen burner flame.
4. Small bits of gold are swept from rocks into the water of the stream and they fall and mix with the mud on the bottom. The gold is heavy and it can be separated from other substances by swirling the mud with some water in a large pan.

# chapter 4
# Elements make compounds

## Introduction

> **Outcomes**
>
> At the end of this chapter you should be able to:
> - State the difference between an element, a molecule and a compound.
> - Identify different compounds and state the common properties of compounds.
> - Make some different compounds.
> - Give examples of common chemical reactions using word equations.
> - State what a chemical change is.
> - Describe the conditions that speed up or slow down a chemical reaction.
>
> Your teacher may give you a copy of these Outcomes for your workbook.

Many useful things in our world are built from simpler parts. For example, your home is built from materials like brick, cement and wood. You can only live in it because the materials have been joined together.

In a similar way, everything in the world is built from elements. When two or more different elements join together, a compound is made. For example, the elements hydrogen and chlorine combine to form the compound we call hydrochloric acid (HCl).

The properties of a compound are different to the properties of its elements. This kind of change is called a **chemical change** or **chemical reaction**. Chemical reactions can be very useful.

Chemical reactions can make new substances, like fibre-optic glass.

Chemical reactions can give us energy.

# Atoms and molecules 4.1

Elements are substances made up of only one kind of particle, or atom. Compounds are made up of two or more different kinds of atoms joined together. The smallest unit of a compound is called a **molecule**.

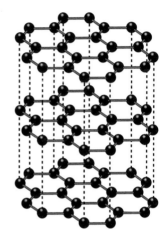

The element carbon.

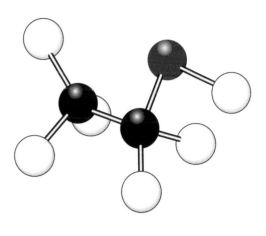

The compound ethanol.

### try this

## Modelling molecules

**COLLECT**
model-making kit

1. Make a model of a water molecule. The **formula** for water is $H_2O$ because each water molecule has two hydrogen atoms (shown by the white balls) joined to one oxygen atom (red ball).
2. Make models of the substances in the following table.

| Substance | Formula |
|---|---|
| Hydrogen | $H_2$ |
| Methane | $CH_4$ |
| Ammonia | $NH_3$ |
| Carbon dioxide | $CO_2$ |

1. Draw and describe an atom.
2. Draw and describe a molecule.
3. Draw two of the molecule models you make. Write the name and symbol beside each one.
4. Imagine you are a water molecule inside a kettle. Describe what happens to you when the kettle is switched on and left to boil dry.

### new words

chemical change • chemical reaction • molecule • formula

**Elements make compounds**

# Making compounds

## 4.2

When a compound is formed, the elements in it are rearranged and cannot be separated again. The compound is a completely different substance to the elements that went into it.

The name of a compound tells you which elements have been joined together to make it. For example, sodium chloride is made up of sodium and chlorine. The colour of a compound can give you a clue to the elements that formed it.

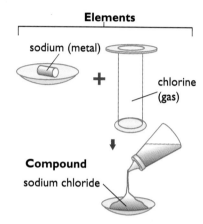

### try this

**COLLECT** samples of the compounds shown below

### Colourful compounds

1. Examine each compound. Try to decide which element caused the colour of the compound. Look for general patterns.

1 Sodium chloride  2 Copper chloride  3 Sodium chromate
4 Potassium permanganate  5 Copper sulfate  6 Potassium chloride
7 Ammonium dichromate  8 Nickel sulfate  9 Potassium chromate
10 Nickel chloride  11 Ammonium sulfate  12 Cobalt sulfate

42  ScienceMoves 2

1. Name the colour sodium is linked with. (Look at compound 1.)
2. Name the colour chloride is linked with. (Look at compounds 1 and 6.)
3. Name the colour copper is linked with. (Look at compounds 2 and 5.)
4. Find other patterns in the colours of these compounds. Report the patterns in a table like the one below.

| Element | Colour |
|---|---|
|  |  |
|  |  |
|  |  |
|  |  |
|  |  |
|  |  |

5. Work out what colour the following compounds will be:
   - cobalt chloride
   - ammonium chloride
   - sodium dichromate.
6. If time permits, have a look at some compounds that contain manganese and iron. Record the colour patterns you see in these compounds.

## Flame tests

1. Soak toothpicks in each of the solutions for as long as possible.
2. Using the tongs, hold a toothpick from one solution in the blue flame of the Bunsen burner.
3. Record the colour of the compound as it burns in the flame.
4. Repeat this for all the solutions.

1. Present your results in a table.
2. Describe any patterns you noticed in this experiment.

**COLLECT**
- solutions of strontium chloride, barium chloride, copper sulfate, sodium sulfate, copper chloride, potassium chloride, lithium chloride, copper carbonate, lithium sulfate, sodium chloride
- toothpicks
- tongs
- Bunsen burner
- heat-proof mat
- safety glasses

### new words
**Add any new words to your glossary.**

# Breaking compounds 4.3

| NP | CSF |
|---|---|
| 4 | |
| 5 | |
| 6 | |
| 7 | |

A compound contains two or more different elements joined together. If the join is broken, the elements may become free. The join can sometimes be broken by electricity.

### try this

**COLLECT**
- large beaker
- connecting leads
- 6 volt DC power supply
- copper chloride
- water
- Bunsen burner
- heat-proof mat
- wooden splints
- gas-collecting equipment
- safety glasses
- metal spatulas for electrodes

## Compound breaker

**Note:** This activity should be performed in a fume cupboard or a well-ventilated area.

❶ Dissolve a spatula-full of copper chloride in about 300 mL of water. Set up the apparatus as shown in the diagram. Switch on the power supply.

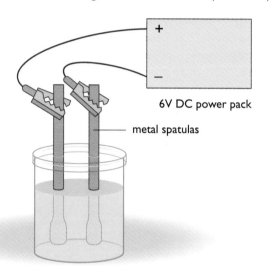

❷ Identify any elements by colour or smell.
**Caution:** Take great care to sniff very gently.

❸ If you have time, repeat the experiment with water in place of the copper chloride solution. Find a way to collect the gases that are produced. Test them with a lighted splint.

1. Describe your experiment(s). Include a drawing of the apparatus, and a description of what you did and what happened.
2. What do you think the elements in the copper chloride are?
3. Write a summary (word equation) of the chemical reaction(s) that occurred in your experiment(s).

### new words

Add any new words to your glossary.

# Rust: an important compound    4.4

Rust is an unwanted compound called **iron oxide**. It forms when the element iron joins with the element oxygen. Rusting can only happen when water is present.

iron + oxygen → rust

| NP | CSF |
|---|---|
| | 4 |
| | 5 |
| | 6 |
| | 7 |

### try this

## Testing rusting

1. Set up the apparatus as shown in the diagram.
2. Leave the experiment for about one week. Observe what happens.

1. Describe whether you think this was a fair test. Describe what you could do to make this a fair test. (**Clue:** Did you set up the apparatus without water to see if rusting still occurred?)
2. What conditions could you alter in this experiment to change the speed and product of the rusting reaction?
3. Design an experiment to test your answer to question 2.

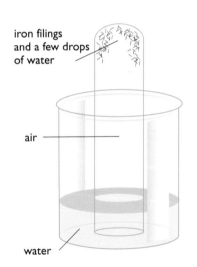

**COLLECT**
- iron filings
- test tube
- beaker
- water

## Rust indicator

Rusting happens quite slowly, but you can show that it is happening by using a **rust indicator**. The rust indicator in this experiment will change colour from pale yellow/green to dark blue/green in the presence of rust.

1. Clean the rust off a nail by rubbing it with sandpaper. Put the nail in some salt water. Add a few drops of rust indicator and observe.
2. Look at the rusty objects. All these objects contain the element iron. They have all been damaged by rust.

    Crumble some of the rust between your fingers. It is orange and flaky, and not very strong.

**COLLECT**
- iron nails
- salt water
- test tubes and rack
- rust indicator (0.1M potassium hexacyanoferrate)
- rusty objects
- sandpaper

**Elements make compounds**

## 4.4 continued...

1. Describe what happens to rust indicator when rust is present.
2. Choose a rusty object.
    a. Name your object and state what it is usually used for.
    b. State how you know that your object is rusting. List your observations.
    c. State whether the rusty object can still be used safely. Explain your answer.

**COLLECT**
- 4 large nails*
- sandpaper
- salt water
- 4 test tubes
- test tube rack
- tongs
- Bunsen burner
- heat-proof mat
- rust indicator
- tin of paint
- expanded polystyrene block
- small beaker of motor oil
- small beaker of plastic powder
- 10 cm lengths of wire
- safety glasses

(*It would be best to prepare the nails a day in advance.)

### Stop the rot

Steel is a metal that contains iron. When iron and steel rust, the metal loses its strength and shape. It eventually becomes useless. There are ways of slowing down rusting, but they can be expensive.

❶ Clean the nails with sandpaper. Prepare one nail by each of the following methods.
   a. Go to the painting area. Dip a nail into paint. Push the nail into a polystyrene block to dry.
   b. Go to the oiling area. Dip a nail into oil.
   c. Go to the plastic-coating area. Heat a nail strongly. Push the hot nail into the plastic powder.

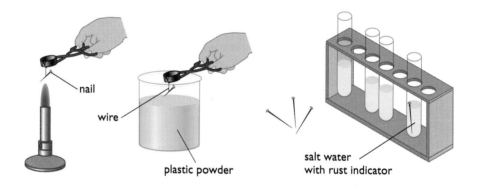

❷ Put each nail in a test tube containing salt water. Add a few drops of rust indicator and observe.

1. Explain why you had to clean the nails first.
2. Record your results in a table with the following headings: Nail covering, Colour of indicator, Rust present or absent.
3. Explain how paint, oil and plastic coating help to slow down rusting.
4. Describe some examples of objects that are protected from rusting by:
   a. paint  b. oil  c. plastic coating.

**new words**

rust • iron oxide • rust indicator

# Chemical reactions 4.5

Chemical changes always produce something new. There are some signs to look for that indicate a chemical change has occurred. These are:
- a colour change
- a gas given off
- a **precipitate** (solid particles) formed
- energy used or given out.

| NP | CSF |
|---|---|
| | 4 |
| | 5 |
| | 6 |
| | 7 |

### try this

## Making chemical reactions (1)

**Caution:** The hydrochloric acid is corrosive.

1. Place 10 drops of lead nitrate in a test tube. Add 10 drops of potassium iodide. Observe.
2. Add 10 drops of hydrochloric acid to a 1–2 cm piece of magnesium ribbon. Observe.
3. Add 10 drops of copper sulfate to a 1–2 cm piece of magnesium ribbon. Observe.

1. In which case(s) was heat produced?
2. Explain which of these experiments involved chemical changes. Give reasons for your answers.

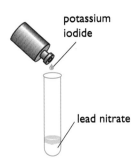

**COLLECT**
- 0.1M lead nitrate in dropping bottle
- 0.1M potassium iodide in dropping bottle
- 0.1M copper sulfate solution in dropping bottle
- 2M hydrochloric acid in dropping bottle
- 2 pieces of magnesium ribbon

## Making chemical reactions (2)

Do as many of these reactions as you can. For each one, decide whether:
a  a new substance is produced
b  there is an energy change.

1. Heat the copper carbonate in a test tube.
2. Add a small amount of hydrochloric acid to a marble chip in a test tube.
3. Shake the magic bottle. Leave it to stand. Shake again.
4. Add zinc to the copper sulfate solution. Stir for a few minutes.
5. Mix the cobalt chloride solution with the sodium carbonate solution.

1. Describe two changes that might happen in a chemical reaction.
2. Write full reports of two of your experiments. Explain how you knew that a chemical reaction was happening.

**COLLECT**
- copper (II) carbonate
- 3 test tubes
- test tube holder
- Bunsen burner
- heat-proof mat
- marble chips
- 1M hydrochloric acid in dropper bottle
- magic bottle*
- piece of granulated zinc
- 0.1M copper sulfate solution
- 0.1M cobalt (II) chloride solution
- 0.1M sodium carbonate solution
- safety glasses
(*The magic bottle contains 6 g of glucose dissolved in 200 mL of 0.5M sodium hydroxide solution in a stoppered conical flask. Add 5 mL of methylene blue.)

Elements make compounds

# 4.5
continued...

## Spot the reaction

1. Discuss the following pictures with your partner. Decide which show chemical reactions and which do not. (Are any new substances made? Is there an energy change?)
2. Write down your answers and the reasons for them.

**a** Mixing water and sand.

**b** Mixing cement.

**c** Using glue.

**d** Boiling water.

**e** Frying an egg.

**f** Burning garden rubbish.

**new words**

precipitate

# Numbers, atoms and formulae    4.6

When you studied the periodic table, you saw that symbols were used instead of the full names of the elements. For compounds, we use **formulae**. The **formula** shows the symbol of each atom in the compound. It also shows the number of atoms in one molecule of the compound. Some examples are shown below.

N–N is $N_2$. N is the symbol for nitrogen and 2 is because there are 2 atoms.

H–O–H is $H_2O$. $H_2$ for 2 hydrogen atoms, O for 1 oxygen atom.

is $CH_4$. C for 1 carbon atom, $H_4$ for 4 hydrogen atoms.

1  Write the formula for each of the following compounds. Set out your answers as in the examples above.

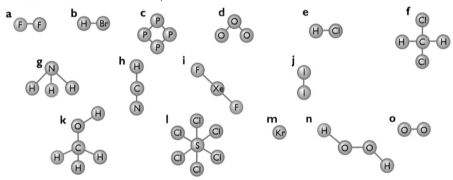

2  For each of the following compounds, list:
   • the names of the elements that make up the compound
   • the number of atoms of each element.

$Na_2SO_3$     KMnO     LiBr     $Na_3AsO_4$
$H_3PO_4$     KOH     $Ca(OH)_2$     $Al_2(SO_4)_3$
$AgCrO_4$     $Ba_3(PO_4)_2$     ZnS     $CCl_4$
$Fe_2O_3$     $Ca(H_2PO_4)_2$     $C_{10}H_{22}$

3  State whether the following are elements or compounds:
   barium sulfate     carbon     copper
   gold     hydrogen sulfate     iodine
   nitric acid     oxygen     silver chloride
   sulfur dioxide     mercury

**new words**
formulae

Elements make compounds  49

# Wonderful chlorine 4.7

| NP | CSF |
|----|-----|
| 4  |     |
| 5  |     |
| 6  |     |
| 7  |     |

Did you know that we could not live without chlorine? We need chlorine ions in our cells, and salt – sodium chloride – is important for maintaining a balance in our bodies. Chlorine is used in industry to make many useful compounds. PVC (polyvinyl chloride) is made from chlorine. This plastic is used to make guttering, pipes, cables, window frames, flooring, car parts, packaging, and the saline drips used in hospitals.

Medicine is another area in which chlorine is important. Cancer therapy products, antiseptics and tranquillisers are made from chlorine. Chloramphenicol is an antibiotic used to treat typhoid, meningitis and pneumonia. Chloroquine is used to treat malaria.

Chlorine is used to disinfect swimming pools and to purify the water we drink. The compound calcium hypochlorite is a bleaching powder used in making paper. Chlorine is found in pesticides and herbicides as well as in perfume bases, sunscreens, food additives, deodorants, cosmetics, detergents, dental cements, flash bulbs ... The list goes on and on.

While many synthetic compounds made with chlorine are extremely useful, it is always important to monitor their effect on the environment. Chlorofluorocarbons are chlorine compounds that have contributed to the destruction of the ozone layer. They are now banned from use. The benefits that new chemicals provide to society must always be balanced against the possible harm they could cause.

### your turn

1. List three compounds made from chlorine and their uses.
2. Write a story called 'The day we ran out of chlorine'.
3. Explain why scientists must take care when producing new compounds and chemicals.
4. Investigate the properties of chlorine that make it such a useful element.

### new words

Add any new words to your glossary.

# Project: Compounds in action 4.8

Your task is to research one compound and to present your research as a 'Science in Action' page.

The following guidelines may help you:
- Describe your compound and the elements in it.
- Explain the uses of the compound.
- Explain whether there are any environmental problems associated with the compound.
- Explain whether we could live without the compound and whether there are alternatives.

Some compounds to look at could be:

acetic acid (vinegar)  
ammonia  
calcium hydroxide (limewater)  
calcium sulfate  
copper sulfate  
nitric acid  
magnesium sulfate (Epsom salts)  
silicon dioxide  

calcium carbonate  
ammonium nitrate  
carbon dioxide  
ethanol  
ethene  
hydrochloric acid  
potassium nitrate  

Your teacher may give you an assessment criteria sheet.

Sodium hydrogen carbonate (also known as baking soda or bicarbonate of soda). It releases carbon dioxide and is used in baking bread and cakes to make them rise. It is also used in some types of fire extinguishers.

## Have I completed everything?

Use the questions below to help you get started, or as a checklist once you think you have finished the research task. This will help you produce a better piece of work.
- Have I read the task outline carefully?
- Do I know what I have to do? (If not, ask your teacher for help.)
- How do I find the answers? Where can I look? What resources can I use?
- How do I organise and present my research?
- Can I add anything extra to improve my work?
- How does this work relate to other work I have done in science? Have I shown these links?
- Is this my best piece of work? Why? Why not?

### new words
**Add any new words to your glossary.**

# another look

## 4.9

### 4.1 Atoms and molecules
1. Describe the difference between an element, a compound, a molecule and an atom.
2. Explain the difference between a physical change and a chemical change.
3. Draw models of three different compounds.

### 4.2 Making compounds
1. Name five compounds and describe their colours.
2. List as many compounds as you can and show how they are related to other compounds by colours.
3. Explain how compounds are made.

### 4.3 Breaking compounds
1. Explain how a compound can be broken, giving an example.

### 4.4 Rust: an important compound
1. Write a word equation to show how rust is formed.
2. Explain why it is important to understand how rust forms.
3. Explain how the rate (speed) of a reaction can be changed.
4. Explain how you could prevent an object rusting.

### 4.5 Chemical reactions
1. Describe what happens in a chemical reaction.
2. List the signs that indicate a chemical reaction has taken place.
3. Briefly describe three chemical reactions that you have observed.

### 4.6 Numbers, atoms and formulae
1. Describe what the formula tells you about a compound.
2. Draw model diagrams of five compounds showing the atoms in them, the names of the elements and the formula of the compound.

### 4.7 Science in Action: Wonderful chlorine
1. Explain why chlorine is important to us.
2. List five compounds made with chlorine and their uses.
3. Explain how science helps us in our everyday lives.

# another look

## Asking questions

It is important to be able to read over your work and ask yourself the kinds of questions you could be asked in a test. Tests often include different types of questions: true/false, multiple choice, fill in the gap, etc.

1. Read over the work in this chapter. Make up:
   - 10 true/false questions
   - 10 multiple choice questions, and
   - 10 fill in the gap questions.
2. Swap your questions with a partner. See if you can answer your partner's questions.

## Crossword

Write clues for this crossword.

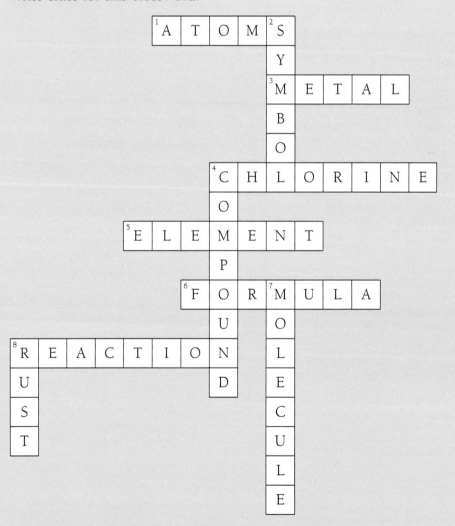

 Your teacher may give you a summary sheet and a topic test.

# Skillbuilder

# Reading skills: finding information

When you have found a passage that seems to contain the information you want, you can scan it to select this information.

The passage below is about materials that are made by combining one or more substances.

1 Scan the passage to find out what each of the following is made from:
   - steel
   - concrete
   - coins
   - glass
   - polycotton.
2 Write your answers in your workbook.

> In the building trade, most materials are made by combining two or more substances. For example, glass is made from sand, soda and chalk. Bricks are held together by mortar, a mixture of cement, sand and water. Concrete is made by mixing sand, cement, water and gravel. It can be reinforced with steel. Steel itself is not a pure substance. It is made from iron with a little carbon added along with tiny amounts of certain other metals such as cobalt. Screws come in many types and sizes. Some are made from brass, which is a combination of copper and zinc. When several different metals are mixed together, the product is called an **alloy**.
>
> The clothes worn on building sites have to be hard-wearing. Some fabrics are synthetic. For example, waterproof coats might be made of PVC, a type of plastic made from carbon, hydrogen and chlorine. More often the clothes will be made from a combination of natural and synthetic fibres. Polycotton is a good example. It is made from cotton and polyester. The coins needed to buy these clothes are made from alloys of copper and nickel.

# chapter 5
# Acids and bases

## Introduction

> **Outcomes**
> At the end of this chapter you should be able to:
> • State the difference between an acid and a base.
> • Describe some common acids and bases.
> • Demonstrate the use of different indicators to identify acids and bases.
> • Solve a problem using the theory presented in the chapter.
> • Make an indicator from plants.
> • State how to neutralise an acid.
> • Describe the uses of some common acids and bases.
> • Design experiments to solve hypotheses.
> Your teacher may give you a copy of these Outcomes for your workbook.

In this chapter you will look at two important groups of compounds that you use every day: acids and bases.

### What are acids and bases?

Water is a compound that contains hydrogen. **Acids** are also compounds that contain hydrogen.

Acids can be dangerous ...        ... or quite safe.

Acids can be cancelled out and made safe by substances called **bases**. Acids and bases are detected with **indicators**. Substances that are neither acids nor bases are called **neutral** substances.

# Investigating acids and bases 5.1

Acids and bases are present even in our soil. Peaty soil (soil that contains lots of decomposing vegetable matter) is **acidic**. Sandy and chalky soil is **basic**. For a long time, people have known that the leaves of some plants change colour in different soils. Hydrangeas, for example, have blue petals when they grow in acid soil and pink petals when they grow in less acid soil.

From this observation, scientists wondered whether it was possible to make a chemical from plants which would indicate whether a substance was an acid or a base. A chemical like this is called an **indicator**. This is the problem you will have to solve at the end of the chapter.

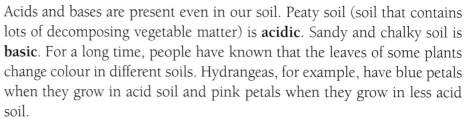

1. Write down the hypothesis or hypotheses you are going to test.
2. Write down all the questions you would need answered about acids, bases and indicators to be able to solve the problem.

You might find the answers to some of your questions in the table below.

| Acids | Bases |
|---|---|
| Acids have hydrogen (H) in their name. | Bases have hydroxide (OH) in the second part of their name. |
| sulfuric acid = hydrogen sulfate ($H_2SO_4$) | ammonia water = ammonia hydroxide ($NH_4OH$) |
| hydrochloric acid = hydrogen chloride (HCl) | limewater = calcium hydroxide ($Ca(OH)_2$) |
| nitric acid = hydrogen nitrate ($HNO_3$) | caustic soda = sodium hydroxide (NaOH) |

A base is the opposite of an acid. If you add the right amounts of a base and an acid together, they will **neutralise** each other.

### Everyday acids and bases

1. Find out what the following acids are used for:
   - acetic acid ($CH_3COOH$)
   - citric acid ($C_6H_8O_7$)
   - the acids listed in the table above.
2. Where would you find these acids in daily life?
3. Find out about where you would find the bases listed in the table above. What are their uses?

**new words**

acid • base • indicator • neutral • acidic • basic • neutralise

# Introducing indicators  5.2

Indicators are used to show whether a substance is an acid or a base. There are many different types of indicators.

**try this**

### Acid or base? (1)

**❶** Test each of the substances with both the blue and the red litmus paper. Copy the table below and complete it by stating what colour change occurred.

| Substance | Red litmus | Blue litmus |
|---|---|---|
| Hydrochloric acid | | |
| Citric acid (lemon juice) | | |
| Sulfuric acid | | |
| Acetic acid (vinegar) | | |
| Water | | |
| Caustic soda (sodium hydroxide) | | |
| Limewater (calcium hydroxide) | | |
| Milk of magnesia (magnesia) | | |
| Ammonia water (ammonia solution) | | |

**COLLECT**
- red and blue **litmus** paper
- 2M hydrochloric acid
- citric acid (lemon juice)
- 2M sulfuric acid
- acetic acid (vinegar)
- water
- caustic soda (sodium hydroxide)
- limewater (calcium hydroxide)
- milk of magnesia (magnesia)
- ammonia water (ammonia solution)
- droppers for the chemicals
- white tile
- universal indicator
- pH paper 1–14

1. Explain why you tested the water.
2. What colour do acids turn red and blue litmus paper?
3. What colour do bases turn red and blue litmus paper?
4. Explain whether litmus paper would be a good indicator to use if you wanted to know how strong each acid or base was.
5. Write a definition of an acid and a base using your answers to questions 2 and 3.

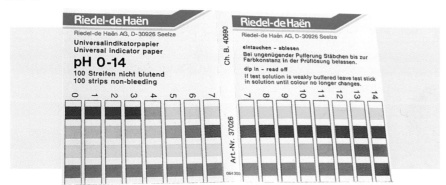

Using **litmus paper** is not the best way to test acids and bases. **Universal indicator** is better. It can come in the form of a liquid or on paper called **pH paper**. The colours of universal indicator are shown opposite. The range of colours is called the **pH scale**.

6. Test the chemicals in the above table again, this time using universal indicator solution and pH paper. Record your results in a table.

# 5.2 continued...

You have just tested some known acids and bases. Now you will use the techniques you have learned to test some common household substances.

## try this

**COLLECT**
- spotting tile
- pieces of pH paper 1–14
- coloured pencils or crayons
- orange juice
- salt solution
- sodium hydrogen carbonate solution
- sugar solution
- egg white
- egg yolk
- coffee
- antacid powder
- cola drink
- detergent
- tea
- vinegar
- different strengths of HCl
- different strengths of NaOH

### Acid or base? (2)

Test each of these everyday substances separately for acidity.

1. Put a drop of the liquid on a spotting tile.
2. Add a piece of pH paper.
3. 
   a. Write the name of the liquid in your book.
   b. Describe the colour that the indicator changes to (or show it using crayon).
   c. Write acid, base or **neutral** beside the name of each liquid.
4. If time permits, try using bromothymol blue and then phenolphthalein on the substances. Record your results in a table.

Think back to the problem with plant indicators that you were given at the beginning of this chapter.

1. State what you have learnt so far to help you solve this problem.
2. Write out how you think you might make your indicator and test it. The pictures below may help you.

**new words**

litmus • universal indicator • pH paper • neutral

# Neutralisation 5.3

An acid can be changed by the chemical reaction of **neutralisation**. In a neutralisation reaction, the acid reacts with a base. When an acid is neutralised, a compound called a **salt** is formed.

acid + base → salt + water (or hydrogen)

Salts are very useful substances. The following are all salts:
- sodium chloride (used as a flavouring and preservative)
- ammonium phosphate (used as fertiliser)
- silver bromide (used in photographic film)
- potassium nitrate (used as a preservative in some meats).

| NP | CSF |
|---|---|
| 4 | |
| 5 | |
| 6 | |
| 7 | |

### try this

### Making copper sulfate

Use the following procedure to make copper sulfate.

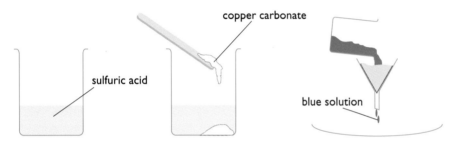

**COLLECT**
- 200 mL sulfuric acid
- copper carbonate
- beaker
- filter funnel
- filter paper
- evaporating dish
- heat-proof mat
- spatula
- stirring rod
- measuring cylinder

❶ Pour the acid into the beaker.
❷ Add the copper carbonate (a base).
❸ Filter the solution into the evaporating dish.
❹ Leave it until crystals form.

1 Describe your experiment using the following words: acid, base, neutralisation.
2 Explain what is happening in each of the cartoons shown here. Try to work out which substance is the acid and which is the base in each case.

### new words
neutralisation • salt

# The fire extinguisher 5.4

| NP | CSF |
|----|-----|
| 4  |     |
| 5  |     |
| 6  |     |
| 7  |     |

Fire extinguishers can be found in every science laboratory. Have you ever wondered how a fire extinguisher works? Do you know how to use one? Did you know you have to tip a fire extinguisher upside down to get it to work? Do you know why?

Acids react with a group of substances called carbonates to produce a salt and carbon dioxide (a gas). The soda-acid fire extinguisher uses this reaction. The extinguisher is full of water mixed with sodium bicarbonate. At the top of the extinguisher is a small vial full of sulfuric acid. When the fire extinguisher is turned upside down, the acid is mixed with the carbonate and a reaction occurs. So much carbon dioxide builds up inside the extinguisher that the pressure increases and forces the water out the nozzle. The water can be directed at a fire to put it out.

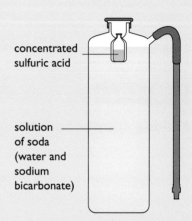

## your turn

1. In your own words, explain how the soda-acid fire extinguisher works.
2. Describe the chemical reaction that occurs in the fire extinguisher.
3. a State the type of fire extinguisher you have in your science laboratory.
   b Describe what kinds of fires you would use this extinguisher for.
4. You have probably observed an acid–carbonate reaction before. Can you give an example of one?
   **Hint:** vinegar.

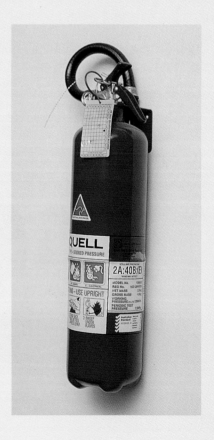

## new words

**Add any new words to your glossary.**

# Acid problems  5.5

**on your own**

### A confusing mix!

Can you solve these problems?

1. Helen and Scott were cleaning up the kitchen and got the jars all mixed up. They knew the jars contained flour, castor sugar, health salts and icing sugar. Help Helen and Scott.

   **Clues:** Flour and icing sugar are insoluble.
   Health salts fizz in water.
   Flour turns a blue-black colour with iodine.
   Castor sugar is the only soluble solid.

2. Carmella and Cameron had a similar problem. Their jars contained cream of tartar, flour, salt and bicarbonate of soda (sodium hydrogen carbonate).

   **Clues:** Cream of tartar is an acid.
   Bicarbonate of soda is a base.
   Flour is the only insoluble substance.

| NP | CSF |
|---|---|
| 4 | |
| 5 | |
| 6 | |
| 7 | |

**try this**

### The dangers of acid

Do you know the effect of acid on body tissue? Do you know how to treat a person if they get acid on themselves? If you don't, make sure you find out.

**COLLECT**
- concentrated sulfuric acid
- dropper
- bull's eye
- dissecting board

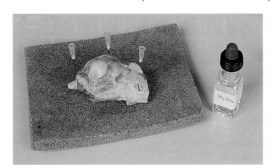

This teacher demonstration involves concentrated sulfuric acid, which is **extremely dangerous**.

1. Your teacher will place a bull's eye on a dissecting board and drop a few drops of sulfuric acid onto the eye and leave it for about 5 minutes.

   1. Describe the effect of the acid on the bull's eye.
   2. Explain what would happen if you splashed acid into your eye? What would you do?

**Acids and bases** 61

# 5.5 continued...

## try this

**COLLECT**
- 500 mL full-fat milk
- water bath at 50°C
- vinegar in a dropping bottle
- lemon juice
- filter paper
- filter funnel
- 2M ammonia solution
- beaker
- several 10 cm squares of card
- retort stand
- thread
- 10 g slotted masses and hanger

### A sticky problem

Milk will react with vinegar, which is an acidic solution, to make a useful glue.

❶ Make some milk glue by following the steps shown in the diagrams below.

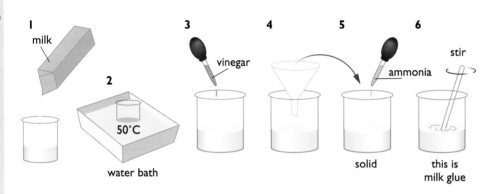

❷ Test the strength of the glue as shown below.

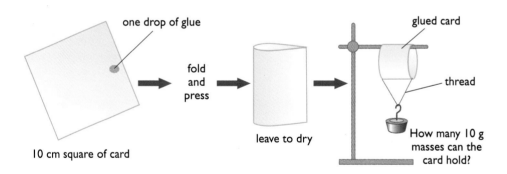

❸ Make another glue by replacing one of the substances in the recipe with lemon juice.
❹ Test the strength of this glue.
❺ Continue until you find a strong glue.

1 Write a report giving your recipe and explaining:
   a  how lemon juice can be used to make glue
   b  how you tested your glue
   c  how strong your glue was compared to the original milk glue.

**new words**

Add any new words to your glossary.

# Reaction riddles                                5.6

Design and carry out an experiment to solve one of the following problems. Write up your experiment as a practical report.

1. Weak acids, such as cola drinks, will clean dirty coins. Do all brands of cola clean coins? Which brand is the best?
2. Bronwyn and Dale knew that antacids, such as Eno or Quick-Eze, neutralise an upset stomach. How could they find out how much antacid was needed to make the stomach environment neutral?

   **Hint:** The acid concentration of the stomach is the same as 1M hydrochloric acid.

3. Kayleigh had made a bottle whiz around her bath by putting bicarbonate of soda (sodium hydrogen carbonate) and vinegar in the bottle. The escaping gas forced the bottle forward. Debbie suggested it might go faster if the vinegar was heated. What could make the bottle move faster?

4. Craig worked in a milkbar making milkshakes. Some kinds of fruit he used caused the milk to curdle. What fruits should Craig use?
5. Cameron and Leigh noticed that their car was rusting. Cameron said it was because they lived near the sea, and the salt spray caused the rusting. Leigh thought it might have been the sand. Who was right?

6. Minh put some Alka Seltzer tablets into a glass jar of vinegar, causing an acid–carbonate reaction. She noticed that if she put mothballs into the jar with the vinegar and Alka Seltzers, they spun and whizzed around. She looked at the size of the bubbles on the mothballs and how fast the mothballs rose to the surface.

   She wondered if it would make any difference to the rate of the reaction if she:
   - altered the amount of Alka Seltzers
   - altered the concentration of vinegar
   - used a different acid, or
   - put in drops of detergent.

   Can you help Minh?

Your teacher may give you an assessment criteria sheet.

# another look

## 5.7

### 5.1 Investigating acids and bases
1. Describe an acid. List some common acids.
2. Describe a base. List some common bases.
3. Describe the uses of one acid and one base.

### 5.2 Introducing indicators
1. Describe the function of an indicator.
2. Name two different types of indicators.
3. Explain why pH paper is a better indicator than litmus paper.
4. Describe the pH scale.
5. Name five acids and five bases and state what colour they change each type of indicator.
6. Describe how you solved the problem presented to you at the beginning of the chapter.

### 5.3 Neutralisation
1. State what neutralisation is.
2. Describe how you would neutralise an acid.
3. Name the end product of a neutralisation reaction. Explain whether the product is harmful.
4. Describe an everyday situation in which neutralisation is important.

### 5.4 Science in Action: The fire extinguisher
1. Describe what happens when an acid comes into contact with a carbonate.
2. Explain how a soda-acid fire extinguisher works.
3. Explain why there are usually two different fire extinguishers in a science lab.

### 5.5 Acid problems
1. Describe how you would test an unknown substance to see if it was an acid or a base.
2. Describe how you would tell the difference between flour and castor sugar.
3. Describe the effect of acid on body tissue. Explain the effects of acid on the eye using the correct terms.
4. Describe how you would make glue from milk.

# another look

## Making concept maps
1. Brainstorm a list of words on the topic of 'acids and bases'.
2. Use your list to draw a concept map. Put the words 'acids' and 'bases' in the centre of a large piece of paper. Arrange the other words around it and show how they are related to acids and bases by making links and explanations.

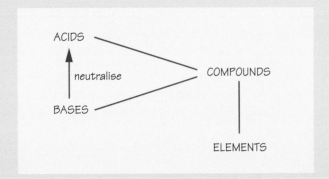

3. Now look back at the other chapters in this section. Choose six key words from each chapter. Using these words, draw a large concept map showing how the three chapters relate to one another.

## Improving your projects

Your project in this chapter was to design an experiment to solve a problem. You should have written up your experiment in a practical report. You will need the report to do this activity.

1. In groups of four, read each other's practical reports.
2. Once you have finished reading, write a list of all the good points in each practical report (for example, one person might have written a thorough method, and included suggestions for further research). List all the positive parts of the reports.
   (**Remember:** In this activity the aim is to learn from each other. Be positive, not critical, when discussing the work of others.)
3. Now look at the marks and the teacher's comments given on these reports. Can you work out why these marks were given?. Why was one experiment better than another?
4. Write yourself a list of things you can do to improve your practical reports.
5. Try this activity with other projects you have been asked to do.

Your teacher may give you a summary sheet and a chapter test.

# Skillbuilder
# Making your own assessment criteria sheets

For the projects you have done so far, you have been provided with assessment criteria sheets. This exercise will help you to develop skills in writing your own criteria.

You will develop assessment criteria for the following project. (Note that you do not have to actually complete the project.)

> **Project**
> Research the different ways we use the chemical reaction of neutralisation in our everyday lives.
> 
> You must find at least 10 examples of neutralisation in the home, workplace and industry.
> 
> Present this information as a poster.
> 
> You must show evidence of your research and provide a bibliography.
> 
> You may like to contact people in different industries to help you with this task.

### Step 1
Work in pairs. Each of you should highlight or underline what you think are the most important words (e.g. research, neutralisation). Compare your list of words with your partner's.

### Step 2
Write down, in one-line statements, what you would need to do to complete the project. For example:
- Research the chemical reaction of neutralisation.

### Step 3
Make sure each of your statements contains only one task. If any statements involve more than one task, break them down into single tasks.

These statements are now your criteria.

### Step 4
Add any other general criteria you consider important. For example:
- The poster must be easy to read.

### Step 5
Decide which are the most important criteria, and which are less important. Then decide how you will assess each criterion. Will you give marks; use a high, medium and low scale; or a well done, OK, poor, very poor scale?

### Step 6
As a class, go through each pair's list of criteria and come up with a list that is acceptable to everyone.

Life and living

section three
# First aid, cells and the environment

science in context

## 6 Life matters

## 7 The bricks of life

## 8 Environments

# science in context
## chapter 6
# Life matters

# Introduction

> **Outcomes**
>
> At the end of this chapter you should be able to:
> - Describe and demonstrate how the lungs work.
> - Describe how the heart works and how the heart and lungs are interconnected.
> - State what first aid is and why it is important.
> - Briefly explain some first aid techniques to save lives.
> - Explain why certain first aid techniques are carried out.
> - Carry out a dissection.
>
> Your teacher may give you a copy of these Outcomes for your workbook.

This section is all about life. In this chapter you will look at life-saving techniques, and how these techniques relate to the heart and lungs in particular. In later chapters you will use a microscope to examine the cells and systems in the human body. You will also look at the importance of the environment in maintaining life.

**Note:** You will need an up-to-date first aid manual to complete this chapter.

Answer the following questions to see how much you know about first aid.
1. What would you do to assist someone who was unconscious?
2. What would you do to assist someone who was not breathing?
3. What would you do to assist someone who had been burned?
4. What would you do to assist someone who was bleeding?

Your teacher may give you a longer first aid quiz.
Your teacher will go through the answers with you.

5. List all the questions about first aid that you would like answered. As you work through the chapter, note any answers to your questions.

# What do I do first?  6.1

**First aid** is the immediate care of a person who is sick or injured. First aid is important because it may stop the sick or injured person (the **casualty**) from getting worse or dying. When you approach a casualty you should follow a set procedure called **DR ABC** (pronounced 'doctor A B C').

| NP | CSF |
|---|---|
|  | 4 |
|  | 5 |
|  | 6 |
|  | 7 |

Danger
Response
Airway
Breathing
Circulation

## Danger

The 'D' stands for **danger**. Before you touch or go near a casualty you should think of the dangers to yourself.

1 Describe the possible dangers in a first aid situation. Think about dangers to yourself as well as to the casualty.
2 Explain how you would assess the dangers to work out how serious the situation was. (**Hint:** Is the casualty awake? Can he/she see, hear, smell? What symptoms does the casualty have?)

## Response

The 'R' stands for the **response** you receive from the casualty.

If there is no response, the person is **unconscious**. An unconscious person cannot protect themselves. If they are lying on their back their tongue can slip down their throat and they can stop breathing. Therefore a person who is unconscious is more important than any other casualty.

1 Describe how you could tell whether someone was conscious.
2 State what you could do to test your idea.

**Life matters**

### Getting help

When you have assessed the situation and the casualty's response, you should send someone for help.

1. If you were with a person who had had an accident, what information about the casualty would you need to give someone so that they could get help for you?
2. Do you have your own list of emergency phone numbers? If not, you should make one. It should include phone numbers of your nearest hospital and casualty section, the fire brigade, the poison information centre, your doctor, the police, the ambulance service, a dentist, a vet, and a chemist. Can you think of any other numbers to include?

### Airway

The 'A' stands for **airway**. If you get no response from the casualty you must act straight away to clear the person's airway to make sure they can breathe. You must put them in a **lateral** (side) **recovery position**.

The lateral recovery position (front view).

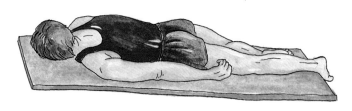

The lateral recovery position (rear view).

1. Using a first aid book, find out about the lateral recovery position. Practise putting your partner in this position.
2. Describe what you must do to a person when they are in this position.
3. Explain why an unconscious person must be placed on the side.
4. State what structure blocks the airway of a person who is not on their side.
5. Explain why we must be able to breathe. Explain what we get from breathing.

**new words**

first aid • casualty • DR ABC • danger • response • unconscious • airway • lateral recovery position

# The lungs and breathing 6.2

While we are still at the 'A' of DR ABC, let's take a closer look at the **lungs**.

Breathing is the action of taking in and letting out air. When we breathe in (**inhale**), air is taken through the **trachea**, or **windpipe**, into the lungs. In the lungs, oxygen from the air passes into the bloodstream and carbon dioxide passes out of the bloodstream. The blood carries the oxygen around the body to the cells that need it. The carbon dioxide is breathed out (**exhaled**).

The air rushes into the lungs when the **diaphragm** is drawn downwards. Air is pushed out of the lungs when the diaphragm relaxes again.

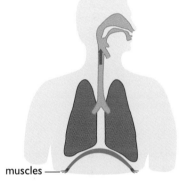
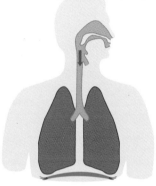
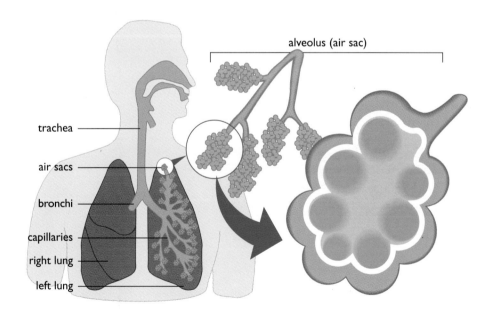

### on your own

**Not just hot air**

Find out more about how the lungs work and how we breathe. The diagrams will help you to get started.

Life matters 71

## The air sacs in the lungs

The lungs are not simply two big balloons that fill up with air. Inside the lungs are lots of tiny **air sacs** that look like bunches of grapes. Because there are so many sacs, they provide much more surface area than a single large 'balloon' would. This surface inside the lungs is where gases pass into and out of the bloodstream.

### try this

**COLLECT**
- ball of string
- 50 cm length of plastic tubing

### Twisted or straight?

1. Without twisting or folding the string, measure how much will fit into the plastic tubing.
2. Now twist and fold the string and measure how much will fit inside the tubing.

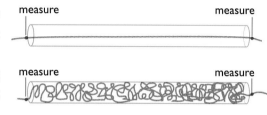

1. Explain why folding increases the surface area.
2. Explain why the lungs are not simply two big balloons.
3. State where else you would find this folding inside the human body. (**Hint:** Think of digestion.)
4. Design another experiment or model to demonstrate this.

### on your own

### Model lungs

1. Design a working model of the lungs. You may need plastic tubing, plastic bags, string, jars, and a 2-litre container.
2. Design an experiment to measure the capacity, or volume, of a person's lungs. (**Hint:** Air can be used to push water out of the way. The diagrams below may help you.)

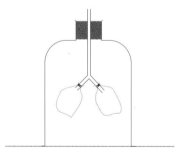
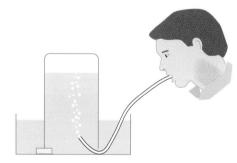

### new words

lungs • inhale • trachea • windpipe • exhale • diaphragm • air sacs

# How to save a life        6.3

### Breathing

Now back to DR ABC. Once you have checked the casualty for danger, response and airway, the next step is to check for breathing. As you have seen, without breathing we could not get oxygen into our bodies, our cells would not receive oxygen and we would die.

You look, listen and feel for breathing by putting your hand in front of the casualty's mouth and placing a hand on their stomach.

### Circulation

If the person is not breathing, the last part of DR ABC is C for circulation.

1. Look at the diagram below. Explain what you should do if you find a casualty who is  **a** unconscious but breathing  **b** unconscious and not breathing.
2. Explain what EAR stands for.
3. Explain what you should do if a person is not breathing but still has a pulse, which means their heart is still pumping.
4. If a person has no pulse and is not breathing, you should carry out CPR. Explain what CPR is.
5. Find out more about these procedures in a first aid book. Spend some time learning what you should do in an emergency.

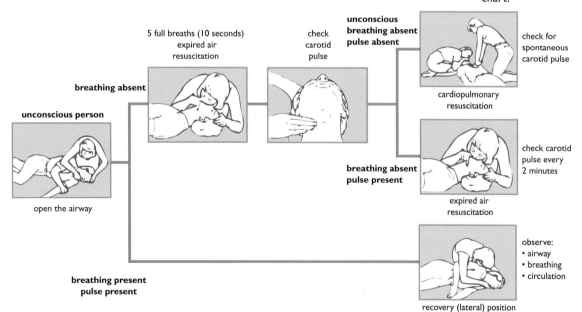

Resuscitation flow chart.

Your teacher may give you a handout about first aid procedures.

**new words**

Add any new words to your glossary.

# The heart 6.4

### How does it work?

The heart is a pump that circulates blood through the body. Your heart beats roughly 100 000 times every day. Over an average lifetime the heart will pump 300 million litres of blood.

As the heart beats, it pumps blood at high pressure through the **arteries** to different parts of the body. From the arteries, the blood enters smaller vessels called **capillaries**. The blood returns to the heart through the **veins**. Veins contain **valves** that prevent the blood flowing back the wrong way. There are also valves in the heart that do the same job.

### What's in the blood?

The blood carries oxygen, which the body needs to stay alive. If the pump stops working, a person can start to die within three minutes because oxygen does not reach the brain.

Blood is made up of plasma, platelets and red and white blood cells. **Plasma** is the liquid part of the blood. It carries the blood cells and wastes from digested food. **Platelets** are the parts of the blood that help it to clot when there is a cut. **Red blood cells** carry oxygen to other cells in the body. **White blood cells** help to fight disease.

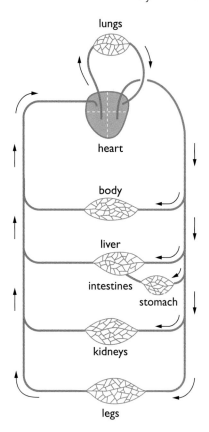

The passage of blood in the middle and lower part of the body.

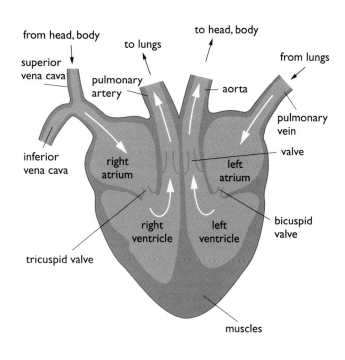

The passage of blood through the heart.

### on your own

### The heart of the matter

Use the diagrams on page 74 and your own research to answer the following questions.

1. State how the blood circulates through the heart and around the body.
2. Explain why the **circulatory system** is connected to the lungs.
3. Describe veins, arteries and capillaries. Explain what job they do and where you find them.
4. Explain what valves are and where you find them.
5. Describe plasma, platelets, and red and white blood cells.
6. Explain the difference between **oxygenated** and **deoxygenated** blood. Explain where you find these.

### try this

### What's inside a heart?

Your teacher may show you the important parts of a sheep's heart.

**COLLECT**
- sheep's heart
- dissection equipment

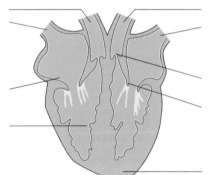

1. Draw a diagram of the sheep's heart and label the parts.
2. Write two sentences about the work that the human heart will do in a lifetime.
3. Describe what a pulse is.

### What's inside a lung?

Your teacher may show you the important parts of a sheep's lungs.
As you watch the dissection, find out:

- what the lungs are like (where they are in the body; their size, colour, texture and protection)
- the path air takes through the lungs (name and describe each part in turn)
- how we breathe (what happens to the ribs and diaphragm).

**COLLECT**
- sheep's lungs
- dissection equipment

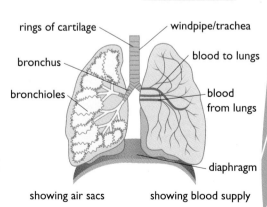

1. In a small group, discuss what you have found out.
2. Write a summary about the lungs. Use the following headings: Description, Path of air, How we breathe.

Your teacher may give you a more detailed worksheet on this dissection.

# 6.4 continued...

You can feel the pumping action of the heart in your arteries. This is called a **pulse**.

## try this

**COLLECT**
- stopwatch
- stethoscope
- dropping bottle of oil
- binocular microscope

### Pulsating reading

❶ Find the pulse in your wrist. Count the beats in 1 minute. This is your pulse rate. If you have a stethoscope, use it to listen to your heart valves opening and closing.

❷ Put a drop of oil on the back of your finger between the nail and the first joint. Look at your finger under a low-power binocular microscope. Try to find the thin red lines of your smallest blood vessels (called capillaries).

1. What was your **pulse rate** (number of beats in 1 minute)?
2. Calculate how many times your heart beats in one hour.
3. You can probably see blood vessels in your wrist. Are these capillaries? Explain your answer.

## on your own

### Heart rate and exercise

Choose one of the following activities to complete.

❶ Design an experiment to test one of the following hypotheses.
- Heart rate increases during exercise.
- The fitter you are, the slower your heart beats when you are not exercising.
- The fitter you are, the quicker your pulse rate returns to normal after exercise.

You will need to consider:
- what variables need to be controlled (such as the age and sex of the subject; and where, when and how the pulse is taken)
- if the activity requires exercise, what exercise will be done, when, where and how
- what will be the control.

Your teacher may give you further work on fitness.

❷ Design a working model of the heart with moving parts. List the materials you will use.

❸ Another circulatory system is the lymphatic system. Research the lymphatic system. Explain how it works and what a lymph node is.

4. For good health, your blood pressure should not be too high or too low. Research **blood pressure**. Explain what it is, how it is measured, and what happens if it is too high or too low.
5. Imagine you are a red blood cell, an aorta, a vena cava, a valve or a ventricle. Write a story describing a typical day. Make it fun and interesting.
6. Design a game that demonstrates how the heart works.
   (**Hint:** You could draw a large diagram of the body and heart on the ground outside. Students could represent different types of blood. (Think of a way to make them look different.) Work out how to demonstrate the blood moving through the heart:
   a  when a person is at rest, and
   b  during exercise.
   Think of a way to show how the blood 'picks up' oxygen in the lungs.

### new words

**the parts of the heart • circulatory system • veins • arteries • capillaries • plasma • platelets • red and white blood cells • oxygenated • deoxygenated • pulse • pulse rate • blood pressure**

## project

# More first aid

You have already been introduced to some basic first aid procedures (DR ABC). However, for most emergency situations you would need to know a lot more about first aid to be able to help the casualty. Many injuries can cause the casualty to lose blood. A major injury, such as breaking the femur (the largest bone in the body) can cause death because of the amount of blood lost.

Your task is to research one of the following types of injury:

| | | |
|---|---|---|
| bleeding | fractured arm | fractured leg |
| fractured rib | poisoning | fractured shoulder or pelvis |
| burns | bites and stings | |

Explain what happens to the body and describe the best procedure to deal with the injury. Demonstrate, using slings, bandages or other medical aids, how you would care for the casualty.

Present your research as a talk to the class.

**Note:** You will need to refer to a current version of a first aid manual such as those published by the St John Ambulance or the Red Cross.

Your teacher may give you an assessment criteria sheet.

# Divers and first aid     6.5

'Diving sickness', 'the bends' and 'decompression sickness' are all names given to a disease that divers, pilots and astronauts can suffer if they experience a drop in pressure too quickly. During a dive, the pressure of the water on the diver's body increases with the depth. At high pressure, nitrogen or helium gas dissolves in the blood. The dissolved gas is carried through the capillaries and into the body tissue.

If the diver surfaces too quickly, there is a sudden change in pressure, from high to low. This causes the dissolved gas to form bubbles in the blood vessels or elsewhere in the body. The bubbles can rupture (burst) or block capillaries, disrupt nerves, and prevent blood flow.

The effects of this sickness include dizziness, loss of vision, nausea and vomiting, tingling in the limbs and paralysis. In the 1900s, people who suffered from 'the bends' drank rum, rested their limbs and hoped to survive. Some did, others were crippled, but most died.

Today, divers can be treated in a hyperbaric chamber – a chamber that can be pressurised. The casualty is subjected to high pressure (2.8 times the normal pressure of the Earth's atmosphere), then slowly brought back to normal pressure over a period of six hours. This allows the dissolved gas to move out of the blood slowly and be exhaled by the lungs.

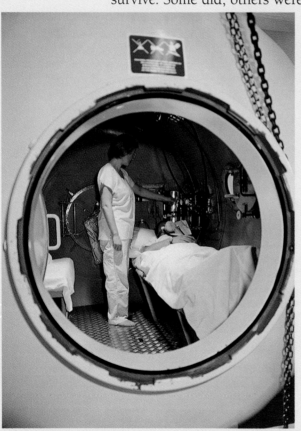

A hyperbaric chamber.

### your turn

1 Describe what happens to divers when they dive into deep water.
2 Explain what happens to divers when they surface too quickly.
3 Describe a hyperbaric chamber.
4 Find out what other diseases are treated using the hyperbaric chamber.
5 Do you have a hyperbaric unit at any hospitals in your area? Find out.

### new words

Add any new words to your glossary.

# another look

## 6.6

### 6.1 What do I do first?

1. State what first aid is.
2. Describe the procedure you must follow if you find a casualty.
3. List the possible dangers you could find at an accident scene.
4. List the ways you could find out about the dangers and what happened to the casualty.
5. Describe how you would test for a response in a casualty.
6. Explain why an unconscious casualty takes priority over any other casualty.
7. Describe what you would do to test whether a person was conscious.
8. When calling for help, what information should you pass on about the casualty?
9. Describe the position you should put a casualty in to protect his or her airway.

### 6.2 The lungs and breathing

1. Explain the job of the lungs and why we need them.
2. Describe how the lungs work.
3. Describe how we breathe in and out.
4. Explain the words 'exhale' and 'inhale'.

### 6.3 How to save a life

1. The procedure to save a life can be remembered by the letters DR ABC. Explain what this stands for. Use a flow diagram to show the whole procedure.
2. Describe what action you would take if a person was not breathing when you arrived at the scene of an accident.
3. Describe what action you would take if a person was not breathing but had a pulse.
4. Describe what action you would take if a person was not breathing and had no pulse.
5. Explain what the letters CPR and EAR stand for.

## another look

**6.6 continued...**

### 6.4 The heart

1. List the arteries and veins connected to the heart.
2. Name the valves in the heart. What does a valve do? Where else do you find valves?
3. Explain why we need valves in the heart.
4. Describe the differences between capillaries, arteries and veins.
5. Describe the jobs of the following: the aorta, the pulmonary artery, the pulmonary vein, the vena cava.
6. Explain why the heart is called a 'double pump'.
7. Describe which side of the heart pumps blood to the lungs and which side pumps blood to the rest of the body.
8. Explain why the left ventricle would have to be thicker and stronger than the right ventricle.
9. Explain why it is more serious to cut an artery than a vein.
10. Describe why the heart and lungs are interlinked.
11. Describe a pulse and the different places you find a pulse in the body.

### Project: More first aid

1. Briefly state how you would treat a first aid problem.

### 6.5 Science in Action: Divers and first aid

1. Explain why divers get 'the bends'.
2. Give a more accurate name for 'the bends'.
3. Describe the treatment for 'the bends'.

23, 24 Your teacher may give you a summary sheet and a chapter test.

# skillbuilder
# Understanding different diagrams

In science you may often find different diagrams of the same thing. The heart is a good example of this. Each reference you use may have a slightly different picture of the heart. Some diagrams may be drawn in a very realistic style, to show what the heart really looks like. Other diagrams may be simplified, to show more clearly how the heart works. These differences can sometimes be confusing. It is an important skill to be able to understand different kinds of diagrams.

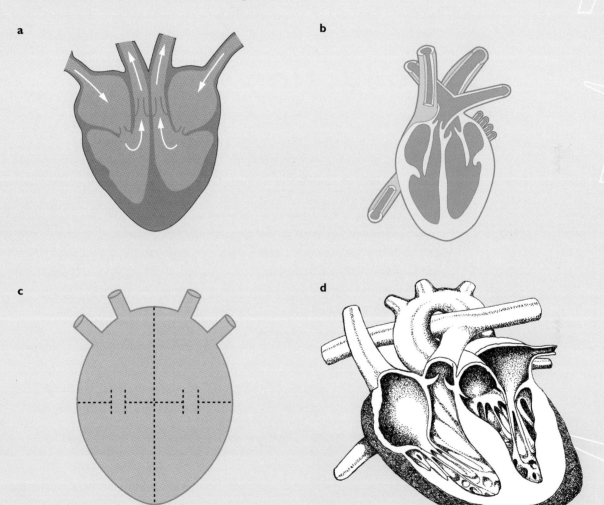

1. Look at the diagrams of the heart on this page. Discuss where you would find the following features of the heart on each diagram: ventricles, aorta, pulmonary artery, pulmonary vein, vena cava and valves.
2. Describe how the four diagrams of the heart are different.
3. State which diagram is the best, giving reasons.
4. Draw your own diagram of the heart.
5. Choose another area of the body and try to find at least three different diagrams of it. Put these drawings on one sheet and label them. Explain which diagram is best and why.

**Life matters**

## chapter 7
# The bricks of life

## Introduction

**Outcomes**

At the end of this unit you should be able to:
- Describe the parts of a microscope.
- Use a microscope correctly.
- State what a cell is.
- Describe the difference between a plant cell and an animal cell.
- Explain how cells contribute to the functioning of organisms.
- State the difference between a cell, a tissue, an organ and a system.
- Describe some different body cells.
- Describe some of the processes carried out by cells.

Your teacher may give you a copy of these Outcomes for your workbook.

In the last chapter you looked at the importance of first aid and investigated how the heart and lungs work. In this chapter you will 'zoom in' to take a closer look inside the bodies of organisms. In the next chapter you will 'zoom out' again to look at how organisms exist in the environment.

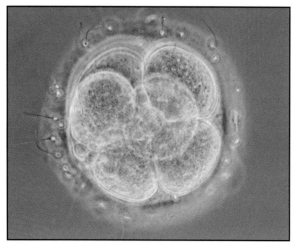

Imagine you looked through a microscope and saw this image. How would you describe what you saw? What would you need to find out to understand what you saw?

### What are organisms made of?

Just as a house can be made up of thousands of bricks joined together, living things are also made up of smaller units. The smallest building blocks of life are called **cells**.

Cells were discovered in 1665 by an English scientist named Robert Hooke. He called them cells because they reminded him of the cells used by monks in monasteries. We now know that all living things are made of cells. The life of a new organism begins with a cell, and organisms grow by making new cells.

# The microscope

## 7.1

We cannot see very small things, such as cells, with our eyes alone – we need help. A **microscope** is a piece of scientific equipment that magnifies small things, making them appear bigger. There are different kinds of microscopes.

Left: A scanning electron microscope.

You are going to use a light microscope like the one shown opposite.

A light microscope magnifies things. Light must be able to pass through the object for you to see it, so the object must be very thin.

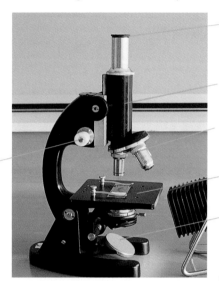

- eyepiece lens (magnifies)
- microscope tube
- objective lenses (magnify)
- focus control (gives you a sharp picture)
- stage (holds the slide)
- mirror (reflects light through your object)

A light microscope.

Your teacher will place a **monocular microscope** and a **binocular microscope** on the front bench.

1. Describe the difference between the two types of microscopes.
2. List some safety rules you think are important when using microscopes.
   (**Hint:** Think about how you should handle the microscope and what substances to keep away from it.) Discuss your rules as a class.

Your teacher may give you a picture of the microscope and some rules to follow.

3. Collect a microscope and take it to your bench. Draw the microscope and label the parts. State what each part does. Do this on the sheet provided or in your workbook.
4. Your teacher will now explain how to **focus** the microscope. Write down each step in your workbook.
5. List the important rules to remember when using a microscope.

**The bricks of life** 83

# 7.1 continued...

### try this

**COLLECT**
• prepared slides

## Using the microscope (1)

Your teacher will give you some prepared slides for this activity.

❶ Collect slide A. Focus the microscope on the slide. Watch what happens to the letter when you move the slide up and down, left and right.
  **a** What is the letter on slide A?
  **b** Describe how the letter moves.

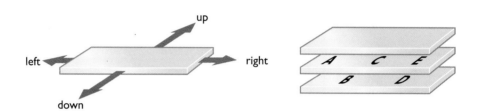

❷ Collect slide set B. There is a five-letter word written on this slide set, but the letters are written on different slides. Focus the microscope up and down to read the letters in order. What is the word?

**COLLECT**
• prepared fibre slide
• clean slide

## Using the microscope (2)

❶ Collect a fibre slide. Use the microscope and the drawings below to identify your fibre. Draw and name your fibre. (See 'Microscope drawings' on page 85.)

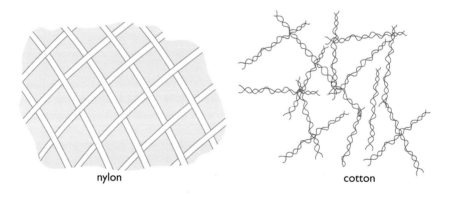

nylon          cotton

❷ Collect a clean slide. Use it to look at some interesting things, such as cloth, hair, paper, fluff, leaves ... Make a large drawing of everything that you look at.

## Magnification

The **magnification** of a microscope tells you how much bigger the image is than the real object. For example, if a microscope has a magnification of × 400 it will show an image that is 400 times bigger than the actual object. (The symbol × stands for 'magnified by'.)

To work out the magnification of a microscope you must look at the **eyepiece lens** and the **objective lens**. Each lens has a number on it, such as × 10, × 20, etc. Multiply the numbers to give the total magnification of the microscope.

eyepiece lens × objective lens = total magnification

**Example**          (× 10) × (× 40) = × 400

Look at the following pictures of a hair at different magnifications.

× 5         × 20         × 100         × 200

1  Describe what happens to the image of the hair as the magnification increases.
2  Describe what happens to the detail of the image as the magnification increases.
3  Describe what happens to the amount of hair you can see as the magnification increases.

## Microscope drawings

You will often need to record what you see under the microscope by making a drawing. When you make a microscope drawing be sure to include the following:

Only draw a small part of what you see. Draw only a few cells in detail.

Date:
Object:
Description:
Stain used (if any):
Magnification:

### new words

**cell • microscope • light microscope • scanning electron microscope • monocular microscope • binocular microscope • magnification • focus • all the parts of a microscope**

# Let's look at cells  7.2

About 300 years ago Robert Hooke made an important discovery using a microscope. You are going to see what he discovered. First you must know how to prepare a microscope slide. Your teacher will show you how.

| NP | CSF |
|----|-----|
| 4  |     |
| 5  |     |
| 6  |     |
| 7  |     |

**try this**

**COLLECT**
- microscope
- slide and cover slip
- dropping bottle of iodine stain
- thin piece of onion skin

### Onions in focus

❶ Make a slide of a thin piece of onion skin.

place thin slice of skin on a microscope slide

add a drop of water

lower a cover slip onto the onion skin

it should look like this

❷ Look at the slide with the microscope using low magnification. Look for a pattern on the slide. What does the pattern look like?

❸ Remove the cover slip and add one drop of iodine **stain** to the slide. (Iodine helps to show the cells more clearly.) Replace the cover slip. Look at the slide again using a higher magnification.

❹ Draw two of the cells. Show as much detail as you can.

### What's inside a cell?

Some very small plants and animals have only one cell, but most plants and animals are made up of many cells. Cells are very small. In one cubic millimetre of human blood there are more than five million cells!

Living cells have different parts which do different jobs.
Most animal and plant cells have:
- a **membrane**, which controls the movement of substances in and out of the cell
- **cytoplasm**, a jelly-like substance that fills the cell; where chemical changes take place
- a **nucleus**, which controls the cell.

Most plant cells also have:
- a **cell wall**, which gives the cell shape and support
- **chloroplasts**, which make food using light
- a **vacuole**, which holds a watery solution.

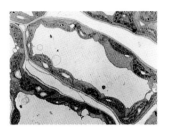
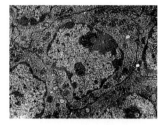

Plant cells from the leaf of a green plant (far left) and animal cells from the inside of a human cheek (left). The main parts of the cells are shown below.

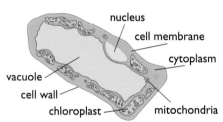
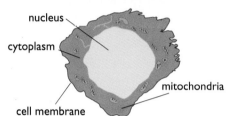

## your turn

Your teacher may give you a diagram of cells.

1. On the diagrams of plant and animal cells, label all the important parts.
2. Copy and complete the table below.

|  | Parts of the cell | What the part does |
|---|---|---|
| Animal cell |  |  |
| Plant cell |  |  |

## try this

### Examining chloroplasts

**Note:** Freshwater plants have thin leaves which makes them good to view. Freshwater plants can be purchased at an aquarium or pet shop.

❶ Prepare a microscope slide as you did for the onion skin, but this time use a small piece of a leaf from a freshwater plant.

❷ Examine the leaf first on low power and then on high power. Draw what you see.

1. Explain whether you would expect an animal cell to look like this.
2. Describe what you would find in a plant cell that is different to an animal cell.
3. Explain why plant and animal cells are different using your knowledge of the parts of cells and their function.

**COLLECT**
- microscope
- slide and cover slip
- leaf from freshwater plant

## new words

**all the parts of plant cells • stain**

The bricks of life  **87**

# Processes inside cells  7.3

As you have seen, the different parts of cells have different functions.

## Getting in and out

Cells need a supply of nutrients, water and gases to carry out their activities. These substances must pass through the membrane of the cell.

Substances such as water move from areas where there is plenty to areas where there is very little water. This is called **diffusion**. Plant cells need a lot of water to fill their vacuoles so that the plant can stand upright.

### try this

**COLLECT**
- agar jelly slab stained with 1.0% phenolphthalein and 0.1M sodium hydroxide
- beaker half filled with 0.1M sulfuric acid
- spoon
- razor blade

### Let me in!

1. Cut the agar into six cubes about 1 $cm^3$.
2. Drop the cubes into the acid. Stir the acid gently to keep the cubes moving.
3. Every 5 minutes remove a cube from the acid. Cut the cube in half.
4. Draw a diagram to show how far the acid has diffused into each cube. (The cube will be colourless wherever the acid has reached.)

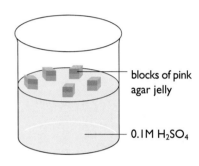

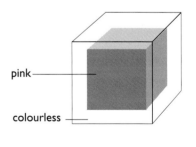

10 minutes

1. Describe how far the acid had reached by the end of the experiment.
2. Explain how this activity demonstrates how water, nutrients and gases get into a cell.
3. Cells need a continuous supply of nutrients. They cannot wait 20 minutes for substances to enter and leave the cell. Explain how the size of a cell helps it receive the nutrients and gases it needs.
4. Explain why the cell membrane is important to the cell.

### on your own

**Jelly test**

❶ Design an experiment using agar jelly to test the following hypotheses:
  - The smaller the cell, the faster diffusion will occur.
  - Diffusion is not efficient in a large organism.

❷ Explain when being large would be an advantage.

❸ For large organisms like humans, explain how nutrients, gases and water get into our bodies and waste is removed. Explain whether humans are too big for diffusion.

### Other cell processes

A lot of activity takes place inside a cell. The activity of a cell is called its **metabolism**. One important process of metabolism is **respiration**. All cells carry out respiration. In respiration, cells use food and oxygen to produce energy and carbon dioxide. This occurs in a part of the cell called the **mitochondrion**.

Another important process for all cells is **reproduction** – making more cells of the same type. When a cell reaches full size the nucleus divides in two. The cytoplasm then divides to form two new cells. This is known as cell **division**.

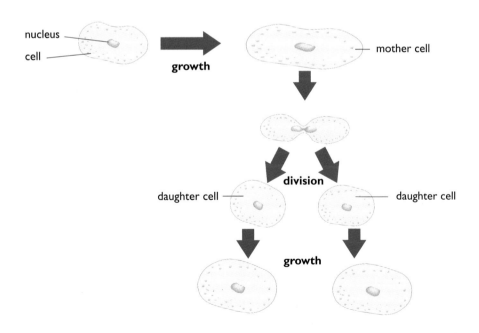

### new words

diffusion • metabolism • respiration • mitochondrion • reproduction • division

# The jobs cells do  7.4

All cells are not the same. They have similar structures but they do different jobs. Cells are said to be **specialised** when they carry out a particular job. Most living things contain different types of cells.

| NP | CSF |
|---|---|
| 4 | |
| 5 | |
| 6 | |
| 7 | |

### try this

**COLLECT**
- microscope slide
- cover slip
- slice of plant stem*
- bottle of stain
- methylene blue

(* Use a microtome for a fine section of stem.)

### Different cells in plants

1. Make a slide of the plant stem. Add a drop of methylene blue.
2. Look at the stem under the microscope using low magnification.
3. Find three different types of cells.
4. Draw the shape of each cell in your book. Write down the job that each cell does. Give your drawing a title.

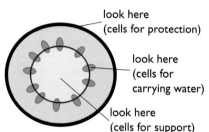

look here (cells for protection)
look here (cells for carrying water)
look here (cells for support)

Your body also has different types of cells. Some of these are shown below.

**Sex cell**
A sex cell carries information. A male makes sex cells called sperm. A female makes sex cells called ova, or eggs.

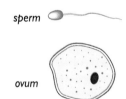

sperm

ovum

**White blood cell**
A white blood cell fights disease.

**Brain cell**
A brain cell passes information through its connections.

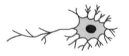

**Motor nerve cell**
A motor nerve cell controls movement. It passes a message from the brain to a muscle.

**Muscle cell**
A muscle cell can make itself shorter to produce movement.

**Skin cell**
A skin cell forms part of a layer which protects the body.

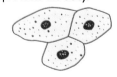

Your teacher may give you a handout about these cells.

1. Name six different types of human body cell.
2. Choose two types of human body cell. Using resources in your classroom and library, find out more about your chosen cells.
   a. List three important facts about each of your chosen cells.
   b. Draw the cells.

1 Look at the diagram below showing some of the different cells in plants. Briefly describe what each cell does.
2 Find the name of the cells that carry water in the stem and the cells that carry food in the stem.
3 Explain why different cells have different jobs.

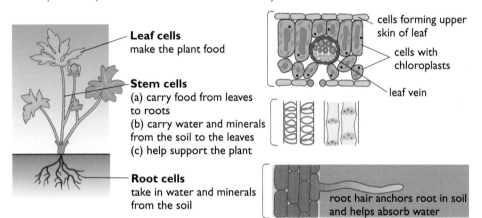

**Leaf cells**
make the plant food

**Stem cells**
(a) carry food from leaves to roots
(b) carry water and minerals from the soil to the leaves
(c) help support the plant

**Root cells**
take in water and minerals from the soil

cells forming upper skin of leaf
cells with chloroplasts
leaf vein
root hair anchors root in soil and helps absorb water

Specialised cells in plants.

## Cells, organs and tissues

1 Look at the diagram opposite. State the difference between a cell, a tissue, an organ and a system. Give examples from the picture.
2 Describe how cells make up functioning organisms.
3 Draw a diagram like the one shown here using the human body as the example.

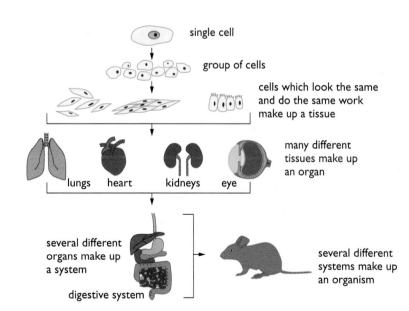

single cell
group of cells
cells which look the same and do the same work make up a tissue
lungs   heart   kidneys   eye
many different tissues make up an organ
several different organs make up a system
digestive system
several different systems make up an organism

Your teacher may give you other diagrams to help you.

**new words**
specialised • phloem • xylem • tissue • organ • system

# Cancer 7.5

| NP | CSF |
|---|---|
| 4 | |
| 5 | |
| 6 | |
| 7 | |

Normally, cells divide at a controlled rate to produce identical new cells that take over the same function. Sometimes this process can go wrong. A 'wild cell' occurs and it divides at an uncontrolled rate. These 'wild cells' don't grow into mature cells that perform their specialised function. Instead, they grow into lumps called **tumours**, which have no function.

The tumours stop healthy cells from working. This disease is called **cancer**. Some tumours are harmless (**benign**). In these tumours, the group of cells is enclosed in a capsule. Warts, cysts and moles are examples of these. If irritated, they can become dangerous.

Other tumours are harmful (**malignant**). Malignant tumours are not enclosed so they spread to other areas where they stop other cells from performing properly.

In developed countries, women are mostly affected by breast cancer and cancer of the uterus. Men suffer more from lung and colon cancer. Both men and women suffer from a high incidence of skin cancer. Digestive tract cancers and rectum cancers seem evenly distributed between the sexes.

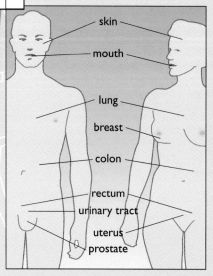

Sites of common cancers in men and women.

1. Cancer has been referred to as 'reproduction gone wrong'. Do you agree or disagree with this statement?
2. Explain whether you would expect lung cancer in women to increase with an increase in the number of women smoking.
3. Explain what affect you think the diminishing ozone layer will have on the incidence of skin cancer.

### on your own

**Research**

Research one form of cancer. State where it occurs in the body and what effect it has. Find out whether it can be cured.

 Your teacher may now ask you to complete a project on the cell (see page 95), or may give you further work on messages in cells.

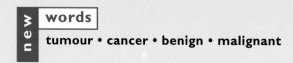

tumour • cancer • benign • malignant

# another look

## 7.6

### 7.1 The microscope

1. Copy the diagram of a microscope and label the parts. Below the diagram, state what each part does.
2. Briefly describe how to focus a microscope on low and high power.
3. List five rules to remember when using a microscope.
4. Describe two different types of microscopes.
5. Explain how you work out the total magnification of a microscope.
6. Explain what happens to the detail of the object being viewed when you increase the magnification of the microscope.

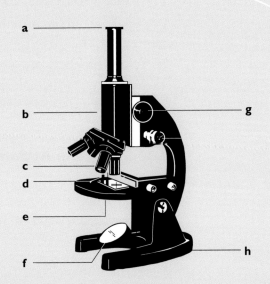

### 7.2 Let's look at cells

1. Describe what onion cells look like under a microscope.
2. Explain why Robert Hooke used the term 'cells'.
3. Explain why you used iodine solution in the experiment with the onion skin.
4. Give the name of another stain that has been mentioned in this chapter.
5. Describe when you would use a stain. Give an example of when a stain would not be used.
6. Draw a diagram of a plant cell showing the different parts. Do the same for an animal cell.
7. State the differences between plant and animal cells. Explain why there are these differences.

### 7.3 Processes inside cells

1. Describe some of the processes that take place inside cells.
2. Define diffusion. Explain how substances get in and out of a cell.
3. Explain why cells must be small in size.
4. Explain how large organisms use diffusion to distribute nutrients, gases and water, as well as to remove their wastes.
5. Explain how you would demonstrate the process of diffusion.
6. Draw a diagram showing how cells divide.

The bricks of life 93

## 7.4 The jobs cells do

1. Name five different body cells and explain the function of each.
2. Explain how a cell's size helps it to obtain nutrients.
3. State what the word 'specialise' means.
4. Name three different plant cells and explain the function of each.
5. Describe how cells make up tissues, organs and systems. Give an example.

## 7.5 Science in Action: Cancer

1. Define cancer.
2. List the most common cancers in men and in women.
3. Describe one cancer, its effect on the body and whether it can be cured.

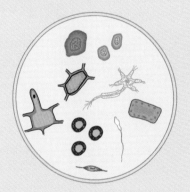

### Mixed up cells

A young scientist was working in her laboratory when she knocked over the sample of cells she was studying. The plant and animal cells were mixed up. This is what she saw when she looked through the microscope.

Can you help the scientist decide which cells are which?

### Linking words

A good skill to have is to be able to get your message across using as few words as possible. Your task is to link the words in the lists below. You must link them in one paragraph of no more than seven sentences. Do this exercise for each of the word lists. You can use the words in any order. Your teacher may give you a summary sheet and chapter test.

| **WORD LIST A** | | |
|---|---|---|
| cell | organism | tissue |
| organ | circulatory system | system |
| heart | lungs | blood cells |
| specialised | functioning | |

| **WORD LIST B** | | |
|---|---|---|
| cell | microscope | specialised |
| uncontrolled | cell reproduction | cancer |
| division | Robert Hooke | monks' cells |

# Reading skills: recognising fact and opinion

A fact is a statement that another person can check. The other person will agree that the fact is true once he or she has checked it. A fact is true for everybody. An example of a fact is:

Dolphins are mammals.

An opinion is a statement that another person might agree or disagree with. It cannot be checked. An example of an opinion is:

Dolphins are cuddly.

Read the sentences below. Decide whether each statement is a fact or an opinion.

1. Science is fun.
2. Scientists are good looking.
3. Chemistry is a science.
4. There are moons orbiting the planet Mars.
5. The rose is a flower that means love.
6. Science fiction novels are more interesting than historical novels.
7. A footballer with a limp is in pain.
8. A tooth is painful when the nerve is affected by tooth decay.
9. Bad teeth are horrible.
10. Smoking can cause heart disease and stomach ulcers.
11. Smoking gives you bad breath.
12. Dolphins are not cuddly.
13. Water is made from two elements called oxygen and hydrogen.
14. A crocodile can grow to 17 metres.
15. My watch is fast.

# project

This project is in two parts. Your teacher may wish you to do one or both parts.

1. Prepare a model of a typical plant or animal cell or a specialised cell. Include all the parts. Your model could be made of cardboard or polystyrene. Use your imagination to come up with creative materials.
2. Write a story titled 'I am a ... cell'. Pretend to be a cell for a day. You can be any cell you wish. Describe yourself and explain what you do and what happens to you in a day.

Your teacher may give you an assessment criteria sheet.

# chapter 8
# Environments

A river bank.

A rocky coastline.

# Introduction

**Outcomes**

At the end of this chapter you should be able to:
- State the difference between an environment, a habitat and an ecosystem.
- Describe the factors and conditions that determine which organisms are found in a particular area.
- Explain the words predator, prey, photosynthesis, herbivore, omnivore and carnivore, and link these in food chains and food webs.
- Describe how living systems are kept in balance.
- Describe some of the cycles in nature.
- State some of the effects of salt on Australian soils and describe what is being done to overcome the problems.
- Examine your local environment using measurement techniques.

Your teacher may give you a copy of these Outcomes for your workbook.

In the last chapter we looked at cells – the smallest building blocks of life. In this chapter we look at the bigger picture – the environment, which plays an important role in maintaining life.

In science, the term **environment** is used to describe all the **conditions** that affect an animal or a plant, such as temperature, rainfall and sunlight. The **place** where an organism lives is called a **habitat**.

1. Describe the habitats and environments shown on this page.
2. List the types of animals and plants you would find in each environment.
3. List the living and non-living things that might affect organisms living in each of these environments.
4. List the problems that organisms would have to cope with in these environments.

# This is my habitat    8.1

For an organism to survive in its habitat, all its needs must be met.

1. Describe what things are needed in a habitat. The pictures below will give you some clues.

2. Describe what factors or conditions could affect a habitat. The pictures below will give you some clues.

3. State which of these are **non-living factors** and which are **living factors**.
4. What have you forgotten? List at least three more examples.

## Describing habitats

A poster or a photograph is a good way to show the appearance of a habitat. Measurements improve the description.

Environments

# 8.1 continued...

## on your own

**COLLECT**
- map of area
- clinometer
- tape measure
- metre ruler
- paper and pens
- camera
- thermometer
- pH probe
- anemometer or wind chart

### Place measurements

1. Discuss how to use the instruments available.
2. Prepare a table like the one below to record your measurements.

| Instrument | What we measured | Our result |
|---|---|---|
|  |  |  |

3. Now go out into the schoolyard and choose a habitat to study.
4. Make and record your measurements.

1. Write a description of the place you studied. Include all the measurements you made.
2. Which of your measurements would change quite often?
3. Which of your measurements would be useful in a weather report?
4. Which of your measurements would be different in six months time?

### Animal and plant numbers

A group of organisms of the same type is referred to as a **population**. There are four factors that affect the number of organisms in a population:

- **birth**
- **death**
- **immigration** (individuals joining the population)
- **emigration** (individuals leaving the population).

Scientists sometimes estimate the size of an animal population by catching a **sample** of the animals in the environment. The sampled animals are always set free afterwards.

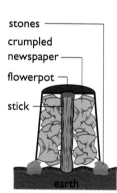
Flowerpot trap

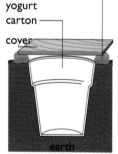
Pitfall trap

Two useful animal traps are the flowerpot trap and the pitfall trap.

## try this

**COLLECT**
- flowerpot
- newspaper
- stick
- yogurt container
- trowel
- tray
- hand lens
- soil organisms identity sheet

### Sampling animals

Your teacher may give you a soil organisms identity sheet.

1. Build animal traps in two different places in the schoolgrounds (or in a nearby park). Hide your traps from bigger animals such as dogs (and people).
2. Leave the traps for 24 hours.
3. Collect the traps and empty them into a large tray. Count the animals.
4. Try to identify some of the animals using the animal identity sheet (or use a key if one is available).

1. Describe the two places where you set your traps.
2. Make a table of results for each place. Use the following headings: Name of animal, Number. If you cannot identify an animal, give it a letter and draw it underneath the table.
3. Choose one animal from each place. Explain why you think it can survive well in that place.
4. Do you think this way of counting animals in an environment is good or poor? Give your reasons.

## Sampling plants

**COLLECT**
- quadrat (square frame measuring 1 m x 1 m)

❶ Find a piece of grassy land.
❷ Choose one type of small plant that seems to be common. (Dandelions, for example, but not grass.)
❸ Place the quadrat on the ground and count the plants inside it. Take three samples in different places.

1. What was the size of your quadrat?
2. Work out the average number of plants in a sample.
3. Estimate the number of plants in 100 m² of this type of land.

### on your own

## Living bottles

Make a 'living bottle'. You can use one of the designs shown here or design your own. You will need a supply of large plastic soft drink bottles.

Your teacher may give you some written instructions.

1. Take any recordings you think are necessary, such as temperature.
2. Describe your living bottle over the course of a few weeks.
3. Explain how these bottles are models of living systems.

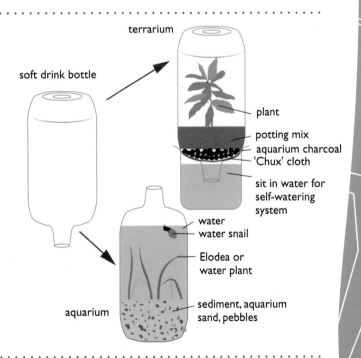

### words
**environment • conditions • place • habitat
• living factors • non-living factors • population • birth • death
• immigration • emigration • sampling**

# Are you dinner? 8.2

| NP | CSF |
|---|---|
| 3 | |
| 4 | |
| 5 | |
| 6 | |

No organism lives alone. When two or more different organisms live together and interact with one another, this is called a **community**. All organisms live in communities. A community of living things together with all the non-living parts of their habitat (air, water, soil) form an **ecosystem**.

## Food chains

Living things in communities are linked together in **food chains**.

All food chains have a similar pattern.

1 Light energy from the sun is trapped by green plants. The light energy is changed into stored energy. This change is called **photosynthesis**.

Plants are called **producers** – they provide nutrients which other animals use.

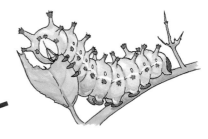

2 The plant uses this energy to grow and stay alive until it is eaten by an animal or dies.

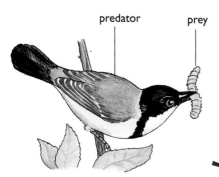

3 The animal uses the energy from the plant. The animal may become the prey of another animal. When it is eaten, energy is passed to the predator.

Animals are called **consumers** – they eat plants or other animals.

4 Organisms that break down dead animals and plants are called **decomposers**.

100  ScienceMoves 2

### try this

**The food chain game**

Your teacher may give you a copy of the food chain game.
Play the food chain game with a partner. The rules are on the game board.

❶ What is the starting point for all food chains?
❷ Explain what the words 'predator' and 'prey' mean.
❸ What is the main change that takes place in photosynthesis?
❹ Which living things carry out photosynthesis?
❺ Why is photosynthesis very important?

**COLLECT**
- food chain game
- 1 die
- red and green pencils

### A sunny start

You are part of many food chains. You eat food. Energy in food comes from a plant or an animal. This energy can always be traced back to the sun.

Write down the food chains for three of the foods on the buffet table.

## 8.2 continued...

### More food chains

Animals that eat plants are called **herbivores**. Animals that eat other animals are called **carnivores**. Animal that eat both plants and other animals are called **omnivores**.

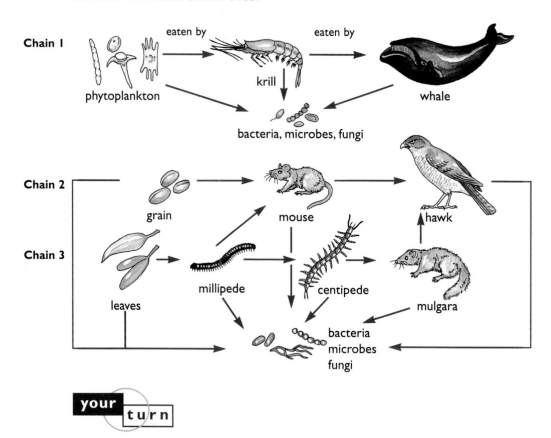

### your turn

1. Copy the above food chains into your workbook (the words only).
2. Put a green circle around each herbivore.
   Put a red circle around each carnivore.
   Put a yellow circle around each omnivore.
3. What is a food chain?
4. Give the meaning of the following words: omnivore, herbivore, carnivore.
5. Would you rather be a herbivore, a carnivore or an omnivore? Give reasons for your answer.
6. What is the real starting point of all food chains?
   (**Hint:** Where do green plants get their energy?)
7. What are producers, consumers and decomposers?
8. List the producers, consumers and decomposers in the diagrams above.

### new words

community • ecosystem • food chain • photosynthesis • producer
• consumer • herbivore • carnivore • omnivore • decomposer
• predator • prey

# Food webs 8.3

Food chains give us an idea of who eats whom, but they are very simple. A more realistic picture of what happens in an ecosystem can be shown in a **food web**. Food webs are made up of a number of food chains linked together. They are very delicately balanced. A small change in one part of the food web could affect many of the organisms within it.

| NP | CSF |
|----|-----|
| 4  |     |
| 5  |     |
| 6  |     |
| 7  |     |

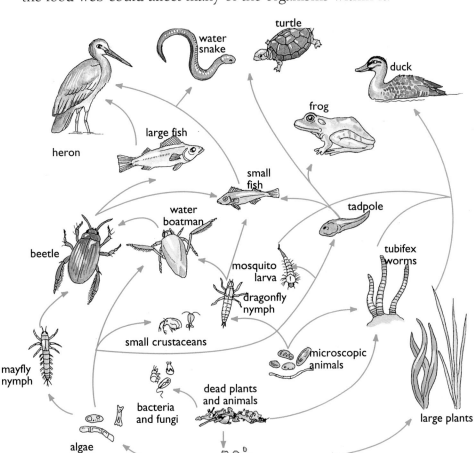

A food web for a freshwater environment.

1. Draw three food chains from the food web above.
2. Name the producers in the food web.
3. Explain how a food web is different from a food chain. Which gives a more realistic picture of an ecosystem?
4. Explain what would happen if you removed the dragonfly nymph from the freshwater environment.

**food web**

Environments 103

# All in balance                                    8.4

| NP | CSF |
|----|-----|
| 3  |     |
| 4  |     |
| 5  |     |
| 6  |     |

In the 'living bottle' activity you carried out, you probably noticed that you had to do very little to keep the models running. Can you explain why?

In ecosystems, **nutrients** are recycled between the living and non-living parts of the system. Plants provide food and oxygen for animals; animals feed off other animals; when the plants and animals die they are broken down by decomposers who return the nutrients to the soil for the plants to use again.

This process is called the **recycling of matter**. Because of it, natural ecosystems are **self-sustaining**; that means they don't need things added to them. There is a balance between all the elements of the system.

It is possible to view the natural environment as a 'web of life'. Each part of the system is connected to the other parts. Like a spider's web, it is very fragile. If we damage one part of this web it affects all the other parts.

Your teacher may give you an activity on worms to demonstrate this concept.

### try this

**COLLECT**
- ball of wool
- scissors

**The web of life**

❶ As a class, find a large open area outside. Ten students stand in a circle and pass the ball of wool around until they are all holding the wool in an unbroken circle.

❷ Other students can then stand inside or outside the circle but they must be connected to the wool.

❸ Now your teacher will cut the wool. If your strand is broken, you must wind your wool up. If you come to another student, he or she must move out of the circle and sit down. Keep winding your wool. What happens?

1. Describe how this model shows the web of life.
2. Explain what happens if part of the web is broken.
3. List ways in which humans break the web of life. List as many examples as you can.
4. Now suggest solutions to at least three of the points you listed above.

## Recycling in living systems

Nitrogen, oxygen, carbon and water are four important elements and compounds that are recycled in living systems.

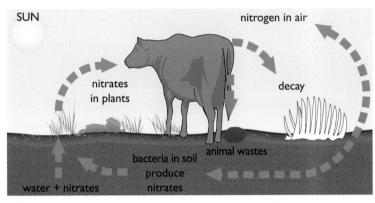

The nitrogen cycle.

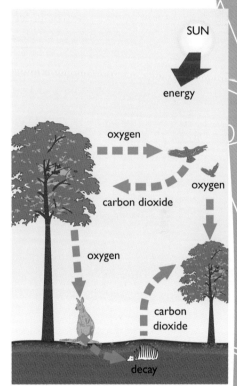

The carbon-oxygen cycle.

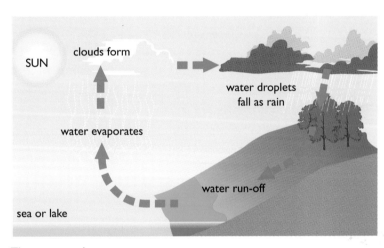

The water cycle.

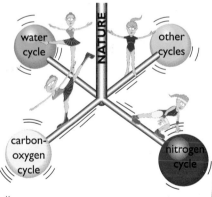

Look at the diagrams above. They show three of the main **cycles** that occur in natural ecosystems.

1. In your own words, explain what is happening in each of the diagrams.
2. Explain why natural systems are self-sustaining. Explain what this means.
3. Explain whether there is a recycling of matter through nature.
4. The last diagram represents the impact humans have had on natural systems. Explain what this diagram means, and how we can correct our mistake.

**new words**

**self-sustaining • nutrients • recycling • cycles**

# Salty Australia

## 8.5

A major environmental problem in Australia is **salinity**. Underneath the top layers of soil in most places there is a layer of **groundwater** – water that has seeped down into the ground and stays there between the grains of soil. This layer of groundwater is known as the **watertable**.

Because many Australian soils contain salt, the watertable also becomes salty. This is not a problem while the watertable remains well below the surface. Plants with deep roots, such as trees, suck up water from the top layers of soil and help to keep the watertable down.

When trees are cleared from the land and are replaced with shallow-rooted crops, less groundwater is absorbed. Any extra water that comes onto the land, such as irrigation water, adds to the watertable and makes it rise closer to the surface. The result is that the top layer of soil in which the crops are growing becomes salty.

Salt is poisonous to most plants. In recent years, many farmers in irrigation areas have found their crop production being reduced. In some cases their crops have died altogether. In Victoria, 96 000 hectares of irrigated land face severe salt problems. Unless the use of this land is carefully controlled, much of it will become useless.

Together, farmers and scientists are trying to overcome the problem of salinity. Some of the methods being tried are:
- replanting native trees
- improving drainage so that less irrigation water seeps down into the watertable
- pumping out groundwater so that the watertable sinks back down.

Land that has been affected by salinity.

1. State some advantages of irrigating dry land.
2. **a** Explain what salinity is and how it occurs.
   **b** Explain what effect salinity has on crops and the land.
3. Describe what is being done to reduce the level of salt in the soil.
4. Imagine you are a farmer who is concerned about salinity. Explain what you will do on your farm to combat the problem.

Your teacher may give you an experiment sheet on the effects of salt on plants and/or more material about pollution and conservation.

### new words

Add any new words to your glossary.

# Investigating pollution    8.6

One way in which humans have an impact on the natural environment is by producing pollution. People often think of pollution as being the smoke that comes from cars and factories. In the following activity you will look at a different type of pollution, known as **thermal pollution**.

You may have seen cooling towers at power stations and steel mills. They cool water down before it is returned to a river or lake. Why is it necessary to do this?

### try this

## Some like it cold

Your task is to find out how warm water could pollute a river by changing the amount of oxygen in it. Marine animals need the oxygen in the water to survive.

1. Set up the apparatus as shown in the diagram. Note the scale reading at the start.
2. Slowly add chemical A from the burette into the beaker containing chemical B. Gently stir the liquid in the beaker. In the correct quantity, chemical A will neutralise chemical B. You will know this has happened when the red colour disappears.
3. When the red colour suddenly disappears, stop adding chemical A. Note the final burette scale reading. (The more of chemical A you needed to add, the more oxygen was present in the water.)
4. Repeat the experiment using warm water instead of cold water.

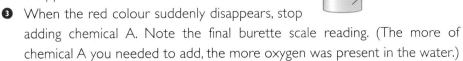

**COLLECT**
- 50 mL burette
- 250 mL beaker
- 100 mL measuring cylinder
- glass stirring rod
- glass filter funnel
- bottle of ferrous sulfate sodium (2.48 g/L), labelled A
- bottle Fehling's B solution (77 g sodium hydroxide, 175 g sodium potassium tartrate, 500 mL water), labelled B
- dropping bottle of phenosaffranin (1% solution, 1 g/100 mL), labelled 'red dye'
- 50 mL cold water

1. What happens to the amount of oxygen in water as its temperature rises?
2. Predict the effect of adding hot water to a river or lake.
3. Explain why adding hot water to a river could be called thermal pollution.

### on your own

## Investigate the local scene

1. Read local newspapers to find out what environmental problems there are in your local area. Report to the class on what you find.
2. Suggest how these problems could be overcome.

Your teacher may give you a worksheet on investigating your local scene.

### new words

**Add any new words to your glossary.**

Environments  107

# Conservation awards 8.7

| NP | CSF |
|----|-----|
| 4  |     |
| 5  |     |
| 6  |     |
| 7  |     |

### Winning an award

You can help to protect the balance of nature in your environment. Take part in the Conservation Award Scheme. Everyone's a winner.

- You will win a certificate.
- The animals and plants in your area will win your care and attention.

1 Complete as many of the tasks as you can, either on your own or as part of a class group. Your teacher will set a time limit.

For each task:
- find out how to do it
- make a plan
- pick the best time and act on your plan.

2 Make an application form for an award. You need to earn:
- 600 points for the gold award
- 350 points for the silver award
- 100 points for the bronze award.

### Fact-finding tasks

|  |  |
|------|--------|
| 1 Find the addresses of four local conservation groups. | **10 points** |
| 2 Write to a local conservation group asking for information. | **10 points** |
| 3 Join a conservation group such as the Australian Conservation Foundation or the Worldwide Fund for Nature. | **50 points** |
| 4 Go around a nature trail. | **50 points** |
| 5 Watch a TV program on nature or read a book about nature. | **10 points** per program or book (up to 100 points) |
| 6 Write a letter to your local councillor asking about her/his ideas on local conservation. | **50 points** (100 points if you get a reply) |
| 7 Collect articles from the local newspaper about local conservation issues. | **10 points** per article (up to 50 points) |

108 ScienceMoves 2

### Fact-giving tasks

| | | |
|---|---|---|
| 8 | Set up a conservation display for the school library. | **100 points** |
| 9 | Organise a talk by a local conservation group at your school. | **100 points** |
| 10 | Organise a visit to a local nature reserve. | **150 points** |
| 11 | Write an article about nature conservation for the school newspaper or for the local newspaper. | **50 points** for school newspaper<br>**150 points** for local newspaper |

### Action tasks

| | | |
|---|---|---|
| 12 | Collect litter from around the school. | **50 points** |
| 13 | Take part in a clean-up campaign in your area. | **100 points** |
| 14 | Make and use a bird table. | **50 points** for each bird table (up to 200 points) |
| 15 | Make a wild garden. You can buy seeds to help you. | **150 points** |
| 16 | Plant a native Australian tree such as a eucalypt with your classmates. Look after it. | **100 points** per tree (up to 300 points) |
| 17 | Set up a tree nursery. (You can do this in a window box.) | **300 points** |
| 18 | Make a nest for insects that live in holes. Use drinking straws or hollow plant stems, or make holes in dead wood. | **50 points** |
| 19 | Raise some money for a local or national conservation group. | **5 points** per dollar |
| 20 | Plan and carry out a survey of wild birds in your area. | **10 points** per type of bird spotted (up to 500 points) |

 Your teacher may provide you with a tally sheet for assessment.

**new words**

Add any new words to your glossary.

# another look

## 8.8

### 8.1 This is my habitat
1 Define 'habitat' and 'environment'.
2 Describe some of the living and non-living things that influence the organisms found in a particular habitat.
3 List living and non-living factors that could affect the environment shown opposite.
4 Define 'population'.
5 Describe two different sampling techniques.

### 8.2 Are you dinner?
1 Describe a food chain.
2 Give the meanings of the following words, and an example of each: predator, prey, omnivore, carnivore, herbivore, producer, consumer, decomposer.
3 Draw a food chain and use each of the words above as a label on your diagram.

### 8.3 Food webs

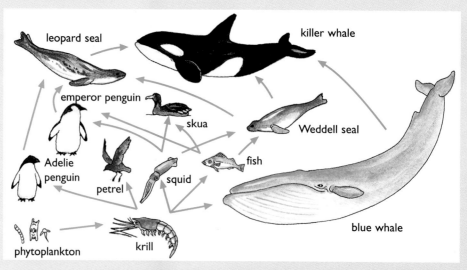

1 How is a food web different from a food chain?
2 Name the producers, consumers and decomposers in the food web shown above.
3 Draw three food chains based on this food web.
4 Some countries fish for krill. Describe what effect this could have on the Antarctic food web.

### 8.4 All in balance
1 Explain why the environment can be described as a 'web of life'.
2 Describe how the natural environment maintains a balance.
3 Describe three different cycles that occur in the natural environment.
4 Explain whether matter is ever lost from a natural environment.

# another look

### 8.5 Science in Action: Salty Australia
1. Describe the problem of salinity.
2. Explain how science is helping to solve the problem.
3. Explain how this problem could be prevented in the first place.

### 8.6 Investigating pollution
1. Define 'thermal pollution'.
2. Explain how thermal pollution could upset the balance in an ecosystem.

### 8.7 Project: Conservation awards
1. List and explain some of the pollution problems you have investigated.
2. Describe some of the conservation strategies you have studied.

## Linking concepts

Look back at the work you have done in chapters 6, 7 and 8. Draw a concept map to show how these chapters link together.

- List six new words from each chapter. An example is shown below.

| Chapter 6 | Chapter 7 | Chapter 8 |
|---|---|---|
| first aid | cell | habitat |
| airway | system | environment |

- Write the words on a large sheet of butchers' paper. Connect the words with lines. On each line write how the words are linked.
- Explain what you know now that you didn't know before you started this section of work.
- Explain why it is important to study the work presented in this section.

Your teacher may give you a summary sheet and a chapter test.

## Strengths and weaknesses

It is important to be able to recognise your strengths and weaknesses when you have completed a topic of work. To do this, answer the following questions:
1. What did I find easy in this section of work?
2. What did I find difficult?
3. Is my workbook up-to-date?

List any problems you may have and discuss them with your teacher.

# Writing skills: using diagrams in notes

A written passage often has a lot of unnecessary details. You should read it to find the main ideas. These main ideas then form the starting point of your notes about the passage. You can note the main ideas in different ways.

> There he lay, a vast red-golden dragon, fast asleep; a thrumming came from his jaws and nostrils, and wisps of smoke, but his fires were low in slumber. Beneath him, under his limbs and his huge coiled tail, and about him on all sides stretching away across the unseen floors, lay countless piles of precious things, gold wrought and unwrought, gems and jewels, and silver red-stained in the ruddy light.
>
> Smaug lay, with wings folded like an immeasurable bat, turned partly on one side, so that the hobbit could see his underparts and his long pale belly crusted with gems and fragments of gold from his long lying on his costly bed.
>
> From *The Hobbit*, by J. R. R. Tolkien (Allen & Unwin 1966)

You might describe the main ideas
- in writing

Big dragon, asleep, wisps of smoke, huge coiled tail, wings folded, lying on piles of precious things.

- or as a diagram.

1  Read the following passage and write a note to describe the main ideas in it.
2  Read it again and draw a labelled diagram to describe the main ideas.

> What is a hobbit? I suppose hobbits need some description nowadays, since they have become rare and shy of the Big People, as they call us. They are (or were) a little people, about half our height, and smaller than the bearded Dwarves. Hobbits have no beards. There is little or no magic about them, except the ordinary everyday sort which helps them to disappear quietly and quickly when large stupid folk like you and me come blundering along, making a noise like elephants which they can hear a mile off. They are inclined to be fat in the stomach; they dress in bright colours (chiefly green and yellow); wear no shoes, because their feet grow natural leathery soles and thick warm brown hair like the stuff on their heads (which is curly); have long clever brown fingers, good-natured faces, and laugh deep fruity laughs (especially after dinner, which they have twice a day when they can get it). Now you know enough to go on with.
>
> From *The Hobbit*, by J. R. R. Tolkien (Allen & Unwin 1966)

The physical world / Energy and change

section four
# Energy, machines and sound

science in context
## 9 Making things move
## 10 Moving machines
## 11 Sounds like ...

# science in context
## chapter 9
# Making things move

## Introduction

> **Outcomes**
> At the end of this chapter you should be able to:
> - Describe what energy is.
> - State the different forms of energy.
> - Identify energy changes in experiments you have performed.
> - Describe stored energy and give examples.
> - Describe conduction and convection.
> - Design your own solar hot water system.
> - Examine colours and materials for energy transfer.
> - Identify how solar energy can be useful to us.
>
> Your teacher may give you a copy of these Outcomes for your workbook.

In this section you will look at **energy**, how energy can be used to make machines work, and finally at sound as a form of energy. Energy can make things work or make things happen. Energy can be **transferred** (passed on) to other things or **transformed** (changed into something different). In this chapter you will look more closely at these different characteristics of energy.

Heat energy.

Sound energy.

Stored energy.

Light energy.

Electrical energy.

Kinetic energy.

# Energy transformations  9.1

1. Draw a concept map of the words 'energy', 'machines' and 'sound'. Add any other words you think are appropriate and draw links between the words to show how they are related.
2. Look at the cartoons on page 114, which show six forms of energy in action. Copy and complete the following table.

| Form of energy | What it can do |
|---|---|
| a Heat energy | Keep me warm |

3. Look at the pictures opposite, labelled **a** to **h**. Write down the main form of energy shown in each picture.

 a Windmill turning.
 e Loudspeaker.
 b Breaking cup.
 f Food.
 c Mains socket.
 g Ice skater.
 d Fire.
 h Object up high.

### Energy transformations

Energy can make things happen or work. Scientists define energy as the capacity to do work. Something only happens when energy changes from one form to another. This is called an **energy transformation**.

For example, a Bunsen burner uses gas. The gas has **stored (chemical) energy**. Burning the gas makes heat. The stored energy of the gas has changed into **heat energy**. Some of the stored energy also changes into **light energy**.

We can write this energy transformation as shown below.

source (input) → receiver → finishing energy (output)

**Example**

gas → Bunsen burner → heat and light

### try this

**Energy changes**

Collect a report sheet. For each experiment below, identify the source (input) of the energy, the receiver and the finishing energy (output). Describe the energy transformation.

1. Hold the thread so that the spiral is above the heater.
2. Make the yo-yo go.
3. Press the switch.
4. Wind up the toy and let it go.
5. Pluck the strings.
6. Switch on the disco strobe.

**COLLECT**
- aluminium spiral attached to thread
- heater
- yo-yo
- switch
- electric bell
- power supply
- alligator clips
- clockwork toy
- guitar or other stringed instrument
- strobe light

| NP | CSF |
|---|---|
| 3 | |
| 4 | |
| 5 | |
| 6 | |

**Making things move**

# 9.1 continued...

| NP | CSF |
|---|---|
| 3 | |
| 4 | |
| 5 | |
| 6 | |

## Stored energy

There are three different types of stored energy: potential energy, chemical energy and nuclear energy.

**Potential energy** is the energy of a stone at the top of a hill or …

**Chemical energy** is the energy in food or …

**Nuclear energy** is the type of energy used in atomic explosions or …

1  Write down three types of stored energy.
2  Copy and complete the table below.

| Type of stored energy | Two examples |
|---|---|
|  | a |
|  | b |
|  | a |
|  | b |
|  | a |
|  | b |

### on your own

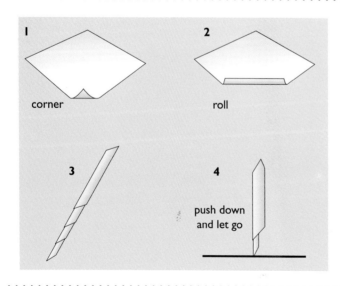

## Paper energy

A piece of paper can be rolled diagonally to make a spring. This spring can store energy.

Use different sizes of the same kind of paper (half a sheet, quarter and so on) and find out:

**a** whether a small spring stores more energy than a large spring.

**b** whether a loose coil stores more energy than a tight coil.

116  ScienceMoves 2

## Difficult energy transformations

Sometimes it is difficult to identify an energy transformation when something happens. This can be because:

1. you start with more than one form of energy
2. one of the forms is difficult to spot
3. the input source of energy is transformed into another form and this form is then transformed again
4. many different energy transformations happen at once.

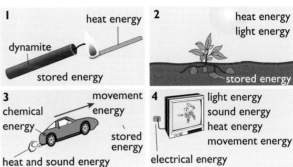

### try this

**Spot the transformation**

Do the following experiments.

1. Set up the apparatus as shown in the diagram. Observe the meter.
2. Dip the rod into one of the chemicals. Hold it in the Bunsen burner flame. Observe the results.
3. Hold the thermometer across your forehead. Look in a mirror.

1. Write a report about one of your experiments. Include the source of energy, the receiver and the energy transformation.

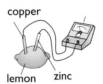

**COLLECT**
**Experiment 1**
- fresh lemon
- connecting wires and alligator clips
- zinc and copper strips
- mA meter

**Experiment 2**
- Bunsen burner
- heat-proof mat
- flame test rods
- lithium carbonate
- copper (II) chloride

**Experiment 3**
- forehead thermometer (crystal strip)
- mirror

### your turn

Identify the most important energy transformation in:

a  a growing plant
b  a moving helicopter
c  film in a camera
d  a lighted candle.

Your teacher may give you handouts on energy.

### new words

energy • transferred • transformed • transformation • stored • chemical • kinetic • potential energy • chemical energy • nuclear energy

Making things move  **117**

# Conduction, convection and radiation     9.2

Heat is energy that transfers from hot objects to cooler ones. There are three main ways that this can happen: conduction, convection and radiation.

## Conduction

If you hold the handle of a teacup and stir the hot tea with a metal spoon, which hand will get hot first? The heat will travel unseen through the spoon and quickly reach your fingers. The transfer of heat energy through materials in this way is called **conduction**.

Conduction usually occurs in solid objects, but not in liquids or gases. Not all solids conduct heat. Those that do not conduct heat are called **insulators**.

## Convection

Another way of transferring heat is by **convection**. Heat can be carried upwards by rising currents of warm gas or liquid. Air expands when it is heated and becomes less dense. Cold air then sinks and forces the warmer air to rise. The moving currents of warm and cold air are called **convection currents**.

**Hot air**
Particles move apart; more volume ⟷ less dense

**Cold air**
Particles close together; less volume ⟷ more dense

### try this

**COLLECT**
- round-bottomed flask
- clamp stand
- washing-up liquid
- spatula
- ice cubes (set with a thread in them so they can be suspended)
- heating coil
- power supply
- candle
- sawdust
- black card

### Convection currents

1. Fill the flask with water. Add two drops of washing-up liquid and a pinch of sawdust. Take care not to breathe in the dust.
2. Gently shake the flask and then allow the water to settle. The dust will show up any convection currents in the water
3. Try each of the following, watching closely each time. You may need to use the black card as a background.
   a  Lower an ice cube into the water.
   b  Hang the heating coil in the water.
   c  Stand a lighted candle below the flask.
   d  Hold your hands against each side of the flask.
4. Draw diagrams of any convection currents you see in your flask.

a  b
water and sawdust dust and washing-up liquid

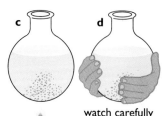
c  d
watch carefully

1. Write a report of your observations from these four activities. Explain what you have seen and felt.
2. Explain whether there can be convection currents in solids.
3. Explain why the cooling unit is positioned in the top of a refrigerator.
4. Explain in which part of a wall you would place an air vent to get the best air circulation.

## Heat watch

Look at the everyday situations below. Make a table to classify them as conduction or convection. Add some other examples of your own.

## Cutting energy costs

The main way that a house loses heat is by conduction through the roof, floor, windows and walls. The picture opposite shows how these losses add up.

Modern houses are built from materials that are designed to reduce these losses and so save us money. The picture below shows examples of how heat losses through windows, roof and walls can be reduced.

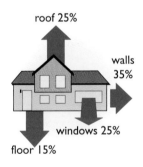

Heat loss from a house.

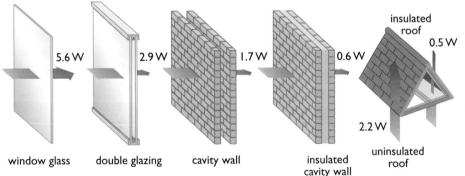

The heat losses shown are for 1 square metre and a 1°C difference between the inside and the outside temperatures.

### Conductor or insulator?

Design an experiment to see which materials are insulators and which materials are conductors of heat.

(**Clue:** Use tin cans wrapped in different materials, such as foam, plastic, material, wood.)

# 9.2 continued...

## Radiation

Not much heat reaches the old man in the picture by convection, because convection currents carry heat upwards. Nor does it reach him by conduction because air is a bad conductor of heat. And yet heat is transferred to him from the heater. How?

The hot bar of the radiator sends out heat waves, or heat **radiation** (also called **infra-red radiation**). Their energy is transferred to the man when they pass onto his skin and clothes.

The Sun's heat energy also reaches us by radiation.

Shiny objects reflect heat radiation while dull objects absorb more heat radiation. Once an object has heated up, it then re-emits its own heat radiation.

### try this

**COLLECT**
- radiant heater (with safety guard)
- copper sheets*
- retort stand and clamp
- thermometer
- timer

(* The copper sheets should be painted black on one side and white on the other. A pocket should be soldered onto each side to hold the thermometer.)

### Light or dark?

Is it better to be a dark-coloured or a light-coloured animal if you want to warm up quickly in the sunshine?

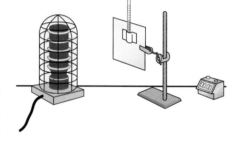

1. Switch on the radiant heater and let it warm up.
2. Put a thermometer in the slot of the copper sheet. Support the sheet on a stand. (It is your 'pretend' animal.)
3. Push the sheet close to the heater and start timing.
4. Measure the temperature of the 'animal' each minute until it heats up.
5. Remove the heater and measure the temperature as it cools.
6. Repeat with the other surface facing the heater.

1. Present your results in a neat table.
2. Plot two graphs of your results on the same temperature/time axes.
3. a What colour skin would help the animal to warm up quickly in the sunshine?
   b Name three animals that have this skin type.
4. Try different colours on your copper sheet. Does the colour affect the temperature of your animal? Test this.
5. Devise a mathematical relationship for the distance from the heat source and the temperature. Devise an experiment to demonstrate it. Try it!

### new words

conduction • insulator • convection • convection current • radiation • infra-red radiation

# Project: Using what I've learned 9.3

Your task is to design a water heater that relies on the Sun and heats water to the highest possible temperature. Look back over the work you have done. Think carefully about the materials and colours you are going to use.

| NP | CSF |
|---|---|
| 3 | |
| 4 | |
| 5 | |
| 6 | |

### Step 1
Decide what materials you are going to use and what colours are appropriate.

### Step 2
Draw a diagram of your heater.
- Are you going to have a control to compare your heater with?
- What are the variables you will need to keep constant?

This picture gives you more information about the properties of heat radiation.

### Step 3
Build your water heater.

### Step 4
Test your design. Record the initial water temperature, then record the temperature at regular intervals. Think of an interesting way to present this information.

### Step 5
Think about how you could modify the design of your heater to improve it. You will then need to retest your new design.

### Step 6
Prepare a talk for your class in which you explain what you did and why. Don't forget to include your results.

Your teacher may give you an assessment criteria sheet.

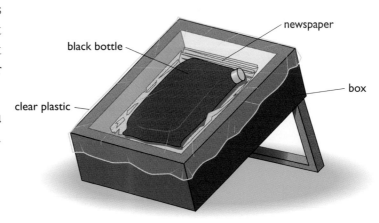

## new words
**Add any new words to your glossary.**

Making things move

# science in action

## The solar cell 9.4

| NP | CSF |
|----|-----|
| 4  |     |
| 5  |     |
| 6  |     |
| 7  |     |

Solar cells can convert light energy into electrical energy. They are also called 'photovoltaic cells' or 'PVs' ('photo' meaning light and 'voltaic' meaning electric). Solar cells are made from silicon, which is extracted from sand and quartz rock. It is the second most abundant element on Earth.

There are several different types of PVs. In mono-crystalline cells the silicon is arranged in a regular lattice pattern. In poly-crystalline cells the silicon is arranged in cube shapes. Amorphous cells don't have a regular crystal structure and are quicker and easier to produce.

Solar cells are used for many things. They are often used in calculators and watches instead of batteries. In California, USA, solar cells are used to help supply electricity for 3500 homes. In remote areas of Australia they are used to provide electricity for railway signals, telephones, water pumps and electric fences. Satellites are powered by solar cells.

Solar cells were originally very expensive to produce and buy. As scientists have developed new ways to manufacture them, they have become less expensive and more efficient. At present, most people still use cheaper alternatives such as fossil fuels to supply their energy needs. However, solar cells will become more important as our supply of fossil fuels runs out.

The use of solar cells is non-polluting because they only rely on the Sun's energy to work. Making the cells does have an impact on the environment, however. Sand mining to produce silicon can damage natural habitats; manufacturing the metal, glass and plastic used in solar cells uses large amounts of energy. We will need to decide whether these effects on the environment are outweighed by the advantages of solar cells over fossil fuels.

1. Explain what PVs are and what these letters stand for.
2. Explain why we don't use solar cells all the time now. Discuss the alternatives.
3. Explain whether solar cells harm the environment.
4. Explain whether you think solar cells are better to use than our present energy sources.
5. Explain the costs and benefits of using solar cells compared with the cost and benefits of using fossil fuels.

**new words**

Add any new words to your glossary.

# another look

## 9.5

### 9.1 Energy transformations

1. Give an example of each of the different types of energy listed below:
   a stored energy
   b sound energy
   c heat energy
   d kinetic energy
   e light energy
   f electrical energy.

2. Copy this table into your workbook.

| Picture | What is happening | Starting energy | Finishing energy |
|---|---|---|---|
| a | | | |
| b | | | |
| c | | | |
| d | | | |

Complete the table using the diagrams below.

3. State the different types of stored energy.
4. Explain why some energy changes are difficult to see. Give examples.

Making things move 123

# another look

**9.5**
continued...

### 9.2 Conduction, convection and radiation

1. Describe the difference between conduction and convection.
2. Describe what convection currents are.
3. Explain how we can use our knowledge of these to build better houses.
4. Describe radiation.
5. Name the best colours to use to reduce the effects of radiation.

### 9.3 Project: Using what I've learned

1. Briefly summarise what you did for your experiment, whether the results were expected and what you found out.

### 9.4 Science in Action: The solar cell

1. Explain how the solar cell has changed our view of energy usage.
2. Explain what role the solar cell will play in providing for our future energy needs.

### Energy word search

Can you find all the words below in the grid? The words run in all directions: horizontally, vertically and diagonally backwards and forwards.

| body | bulb | cell | watt |
|---|---|---|---|
| coal | dynamo | electrical | engines |
| fire | food | fuel | grow |
| heat | heater | hydro | kilojoule |
| light | machine | movement | music |
| oil | plug | power | solar |
| sound | stored | switch | |

```
B Q T E L U O J O L I K B
O S N L Z R A L O S A M U
R H E A T L E C I P F D L
D P M O H S E V D L H Y B
Y A E C G R E W A U C N Q
H D V R I N N C R G T A S
R W O F L A I M A R I M O
E W M B T R H M D E W O U
T A F U T O C F H W S V N
A T U C S J A S T O R E D
E T E G H I M Z T P S O E
H L L E C I C U Y L O I G
E N G I N E S T B F I L U
```

Your teacher may give you a copy of the word search, a summary sheet and chapter test.

# Writing skills: using diagrams to shorten notes

In a well-written passage each paragraph tells you about one main idea. The paragraph may also give:
- facts about the main idea
- examples of the main idea
- opinions of the main idea
- explanations of the main idea.

A spider diagram is a good way of organising the information you read about. It is a good starting point for making notes.

The passage below explains some of the problems of making a human-like robot called an android. Read the passage to find three major problems with designing an android.

> The idea of making a machine that behaves like a human – an android – is fascinating.
>
> A vast amount of work has been done, but at the end of it all we have to admit that compared with Mother Nature, we know very little about either computing or engineering. Let us look at the major problems one at a time.
>
> We need a machine that is capable of moving itself about over rough ground, up stairs, up trees, and even cliffs for several days before its batteries run down. It should be able to pick up the equivalent of its own weight and stagger with it. The same hands and arms must be capable of picking up and threading a needle. We have nothing remotely capable of this sort of performance. A machine using electric motors and batteries would go on the flat, for less than an hour before running down – one flight of stairs would exhaust it. A mechanical arm strong enough to use for a tug-of-war would weigh nearly 45 kg. Even if we knew how to build walking legs (which we do not yet), they would come out weighing more in the region of tonnes than kilograms.
>
> The human eye has the equivalent of some three million pixels; the best available televisions have 1 million. But even if we had a suitably sensitive mechanical eye, we could not process its information in less than hours – as against the eye and brain's $1/25$ second – and we still do not know how to do more than 1 per cent of the necessary processing.
>
> The human brain contains 10 000 million neurons. Each neuron is attached to many others and has an unknown value in terms of byte storage, but we might assume that it can store 100 bytes. The brain is then equivalent to a million million bytes or the contents of a 30-metre cube full of today's memory chips. And that again assumes that we would know how to organise the memory if we had it. One of today's 16-bit micros would take more than three weeks to search that much memory for one four-letter word.
>
> From *The Joy of Computers*, by P. Laurie (Hutchinson 1983)

1   Copy and complete the spider diagram below.

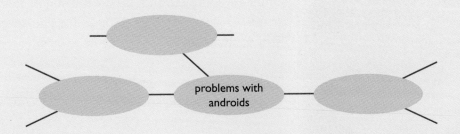

2   Read each paragraph again. Add two explanations for each of the problems you have identified.

**Making things move**  125

## chapter 10
# Moving machines

## Introduction

**Outcomes**

At the end of this chapter you should be able to:
- Describe what a machine is and state why we use machines.
- Make your own simple machine.
- Recognise different simple machines.
- Examine more complex machines and explain how they work.

Your teacher may give you a copy of these Outcomes for your workbook.

In the last chapter you looked at different forms of energy and how they can be used. In this chapter you will looks at energy transformations that make machines work.

## Your machine

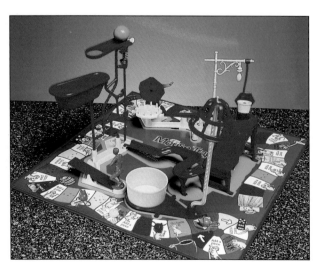

Mousetrap game © Milton Bradley

Your task in completing this chapter is to design and build your own machine using as many simple machines as you can. Write a list of the questions you need answered in order to complete this task. (You might like to play a game such as 'Mouse Trap' to give you some ideas.)

Your teacher may give you an assessment criteria sheet.

# What is a machine?

## 10.1

**Machines** use energy to get a job done. They reduce the amount of **work** required to get a job done. Work happens when a force is applied to an object.

Machines are also energy transformers. They transform one form of energy into another more useful form. The energy output (finishing energy) is different to the energy input (source energy).

| NP | CSF |
|---|---|
|  | 4 |
|  | 5 |
| 6 |  |
| 7 |  |

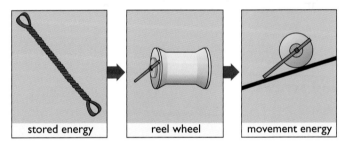

### try this

### The reel wheel

Your teacher may give you a worksheet for this activity.

The picture below shows a machine called a 'reel wheel'. It changes the stored energy in a rubber band into kinetic (movement) energy.

**COLLECT**
- plastic or wooden cotton reels
- candle
- piece of candle with hole bored through
- rubber band
- nail

❶ Make a reel wheel using the picture as a guide.
❷ Race your wheel against a few others in the class.

1. Describe a machine. Give two examples from this page. Write down the energy transformation for each example.
2. Write a full report about making the reel wheel. Include the following:
   - name of machine
   - how the machine works
   - the main energy transformation in the machine.
3. For each machine on this page, state the source energy (input) and the finishing energy (output). Also state the name of the receiver of the finishing energy.

### new words

**machine • work**

Moving machines

# Different types of simple machines    10.2

## Inclined planes and pulleys

Two simple machines are the **inclined plane** and the **pulley**. An inclined plane is another way of saying a **ramp**. A pulley is a rotating disk with a grooved edge. These machines can help reduce the amount of work needed to do things.

### try this

**COLLECT**
- spring balance (scale in Newtons)
- retort stand and clamp
- smooth string
- block of wood with hook attached
- board for ramp
- pulley

### Testing machines

❶ Carry out the experiments shown in the diagrams below.

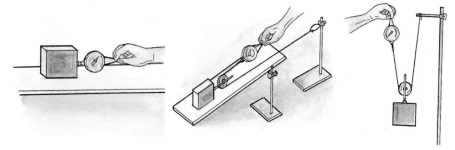

❷ Alter the weight of the block (the **load**) and the height of the ramp. See what effect this has on your force measurements.

1 Explain how **a** a ramp and **b** a pulley reduces the work needed to pull a block.

2 Explain what happened when you changed the load.

## Wedges and screws

A **wedge** is like an inclined plane, but the wedge moves into an object, as in an axe cutting into a tree.

1 Explain how the wedges shown opposite move into objects.

A wood-splitting wedge, an axe and a golfing wedge.

A **screw** can be thought of as an inclined plane that has been wrapped around a cylinder. A screw can penetrate objects.

128 ScienceMoves 2

1. Make your own screw from a piece of paper.
2. Describe what these objects penetrate and how they reduce the work required.

A fan, a car jack, a nutcracker, a corkscrew and a G–clamp.

## Gears

A **gear** wheel is a wheel with teeth around it. Gear wheels usually work together. One gear wheel drives the other to do the work.

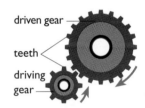

Gearing down – the small gear drives the large gear. The large gear turns slowly.

1. Look at the diagrams opposite. Which gear is making the other one work? What is each gear called?
2. Pairs of gears can be used to 'gear up' or 'gear down'. For example, to reduce speed a small gear is used to drive a big one. The big one turns slowly. Anything attached to it also turns at a slow speed. This is called 'gearing down'.
   Draw examples of 'gearing down' and 'gearing up'.

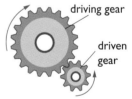

Gearing up – the large gear drives the small gear. The small gear turns very quickly.

## Wheels and axles

1. Look at the pictures shown below. Explain how a **wheel** and **axle** help reduce the force needed to carry out a task. You may like to build one of these to test your idea.
2. Does the size of the wheel make any difference to the force required?

**Moving machines** 129

# 10.2 continued...

## Levers

Imagine trying to build an Egyptian pyramid in the schoolground. You would not be strong enough to produce the force necessary to lift a single stone. This does not mean that you could not do it. All you would need is a **lever** and a **pivot** (or **fulcrum**).

A lever seems to allow us to increase the force we can produce. This idea of multiplying force is used in many simple machines.

### your turn

Look at the diagrams below.

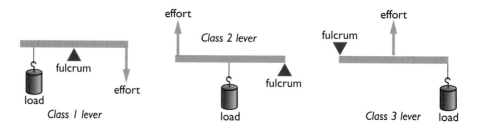

1. Describe the three types of levers presented here.
2. Give examples of the three different types of levers using everyday machines. (The pictures on this page may help you.) Present your information in a table.

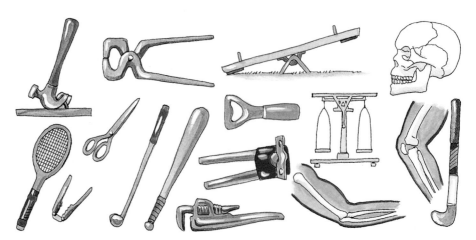

You can use a see-saw to learn more about how a lever works.

### try this

**Balancing act**

**COLLECT**
- 7 small masses (such as five cent coins)
- lever (ruler)
- pivot (pencil or dowel rod)
- rubber band

❶ Set up your see-saw. Use the elastic band to hold the ruler in place.

❷ Experiment with the lever and pivot to solve these balancing problems. There is usually more than one solution. Place the masses on the whole number lines on the ruler. You can place the pivot wherever you like.
Try to balance the masses in piles as follows:
- 4 on the left of the pivot, 3 on the right
- 5 on the left of the pivot, 2 on the right
- 6 on the left of the pivot, 1 on the right.

1 Give three examples of a lever in action.
2 Copy and complete the table below to show the results of your experiments.

| Left side | | Right side | |
| --- | --- | --- | --- |
| Number of masses | Distance from pivot (in cm) | Number of masses | Distance from pivot (in cm) |
| 4 | | 3 | |
| 5 | | 2 | |
| 6 | | 1 | |

3 Explain whether there is a pattern in your results.
(**Hint:** Multiply the mass by the distance on each side of the pivot.)
4 Describe where you would place a 2 g mass to balance the see-saw if the other side had:
  **a** a 10 g mass 2 cm from the pivot
  **b** a 6 g mass 6 cm from the pivot.

### new words

**inclined plane • pulley • ramp • load • wedge • screw • gear • wheel • axle • lever • pivot • fulcrum**

# Examining machines 10.3

| NP | CSF |
|---|---|
| 4 | |
| 5 | |
| 6 | |
| 7 | |

Machines are useful pieces of equipment that help us carry out work. Some machines are simple, such as a door handle, and others are complex, such as a sewing machine or a car.

**Simple machines** have only one type of function, or work on only one principle; **complex machines** are combinations of simple machines. Machines are all around you: in the house, in the garage, at school and in industry.

### try this

**COLLECT**
- corkscrew
- screw
- golf club
- axe or tomahawk
- toy car
- pizza cutter
- bottle opener
- garlic press
- ramp for toy cars
- hand drill
- egg beaters
- Lego Technic

## Which machine?

❶ Look at the collection of machines. Sort them into the various types of simple machines you have studied so far. Write down the names of the machines and fill in the information about them in a table.

| Machine | Machine type | Where found | What it does |
|---|---|---|---|
| | wheel and axle | | |
| | inclined plane | | |
| | lever | | |
| | wedge | | |
| | pulley | | |
| | gear | | |
| | screw | | |

❷ Make one of each type of machine with the Lego kit. Describe how it works.

**COLLECT**
- different types of egg beaters

## Gears and cogs

Your teacher may give you a worksheet for this activity.

❶ Look at the various egg beaters. Are they all the same? List their similarities and differences.

❷ Look closely at the moving parts of an egg beater. Fill in the information on your worksheet.

  1  State whether the blades move faster or slower than the handle.
  2  Explain how the movement of the handle turns the blades.
  3  State whether the larger cogwheel turns faster or slower than the smaller one.
  4  State the mathematical relationship between the number of cogs on each wheel and the number of turns it makes.
  5  Describe the advantage of the egg beater over a whisk or a fork.

Your teacher may give you additional worksheets on machines.

### new words
**simple machine • complex machine**

132  ScienceMoves 2

# Bigger machines

## 10.4

| NP | CSF |
|---|---|
| | 4 |
| | 5 |
| | 6 |
| | 7 |

**try this**

### Steam machine

**COLLECT**
- large machines on display

Your teacher will show you a big machine; perhaps a steam engine.

1. Discuss how the machine works.
2. Describe the important energy transformations that occur when the machine is working. (There are at least three energy transformations.)

Your teacher may give you some instuction cards to assist with this activity.

3. Go to one of the machines set up in the room.
4. Read the instruction card carefully.
5. Make the machine work and study it.

1 Write a report about each of the machines you studied. You should include sentences about:
   - how the machine works
   - the important energy transformations that take place when it is working.

2 Copy and complete the table below for the machines labelled **a–d**.

| Picture | Name of machine | Important energy transformation |
|---|---|---|
| a | | |

a

b

c

d

3 Some machines waste energy. A car changes stored energy into movement energy. However, in the process energy is wasted as heat and sound. Explain whether any energy is wasted in the machines you just looked at.

4 Explain whether energy is ever lost. Can we always identify what energy has changed into?

**new words**

Add any new words to your glossary.

# science in action

# The Internet  10.5

| NP | CSF |
|---|---|
| 4 | |
| 5 | |
| 6 | |
| 7 | |

Computers are very useful machines that can store and rearrange information. Computers can also be linked to other computers all over the world, so a huge amount of information is available to us.

The enormous web of connected computers is called the Internet. The 'information superhighways' that connect the computers are actually just thousands of kilometres of cables, including telephone cables, that run all around the world.

Originally most of the cables were made of copper. Anything can be sent along these cables as long as it can be converted into electrical signals. Sound can be converted into electrical signals and sent along these cables. That is how the telephone works. Fax machines convert text and pictures into electrical signals so they can be sent along these cables.

Today, however, new cables are being made from optical fibres, which carry signals as light rather than electricity. Light travels much faster than sound so this is a more efficient way of getting messages from one place to another. Most major towns and cities in Australia are linked by optical fibre cables and in the future our homes may be too.

Because of these links, you can use information from all over the world. If you want to plan a holiday to London and you need to know train timetables, you can ask your computer to search the Internet to find the London Underground. On your computer screen you will see the right train to take for your journey and all the stops along the way.

Information on the London Underground, available via the Internet.

1. Make a list of all the things that computers help us with in our daily lives. Can all these things be described as 'storing and rearranging information'?
2. Explain how the Internet could be used at your school.
3. List three topics you have covered in science that you would like to research using the Internet. Identify the key words you would use in your search.
4. If you have access to the Internet at school, carry out your research after consulting with your teacher.
5. The Internet is an example of new technology that is changing people's lives. Predict what you think the future will be like in 50 years time. Write about your ideas in a story called 'Trip to the Future'.

**Add any new words to your glossary.**

# another look

## 10.6

### 10.1 What is a machine?
1. Define the words 'machine' and 'work'.
2. Describe the relationship between machines and work.
3. Explain how machines convert energy from one form to another. Give an example.

### 10.2 Different types of simple machines
1. Name the different types of simple machines you have studied. State what each one looks like and how it is useful.
2. Explain the difference between 'gearing up' and 'gearing down'.
3. Look at the diagrams below and explain how levers are used in the design of each machine.

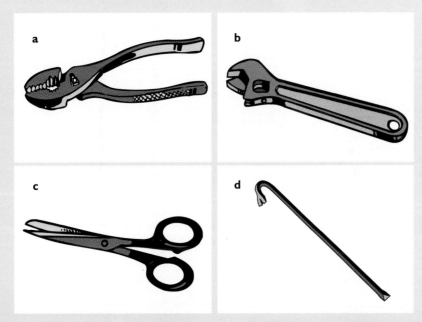

4. State which class of lever is used in each machine.
5. Give three examples of how levers help reduce the work needed to carry out a task.
6. Describe some everyday examples of the machines mentioned in this chapter.

### 10.3 Examining machines
1. Describe some of the machines you examined in this section.
2. Explain how an egg beater works.

## 10.4 Bigger machines

1  Briefly describe three of the bigger machines you studied.
2  Choose a complex machine and state how many simple machines it is made of.
3  Choose three machines and state what energy transformations happen in the machine.
4  Explain whether any of the machines you studied waste energy.
5  Explain whether any machines lost energy.
6  Choose one machine and state what energy input is required to run it.
7  Look at the machine in the picture below.

  a  List the forms of energy that are used in this machine.
  b  Explain whether energy is wasted in this machine.
  c  Explain whether this machine could go on forever.

## 10.5 Science in Action: The Internet

1  Describe the Internet.
2  Explain how the Internet works.
3  Briefly describe any research you did using the Internet.

 Your teacher may give you a summary sheet and a topic test.

# Skillbuilder
## Writing skills: recording facts in notes

Many science books contain good ideas and lots of interesting facts. You might want to understand the ideas. You might also want to remember some of the facts.

To record facts properly:
- Skim the passage.
- Write down the main ideas from the passage (only two or three at any one time). Leave several lines between each main idea.
- Read the passage again. Write each fact that you want to remember underneath the correct main idea.

1 Read the passage below. It has two main ideas:
   - the first computers
   - how computers have changed.

   Write these headings in your book leaving four lines between each one.
2 Read the passage again and write three interesting facts underneath each of the correct main ideas.

> Charles Babbage was an English mathematician. He was born in 1791 and died in 1871. Amazingly he designed the first true computer – nearly 200 years ago! Babbage realised that if he could make a machine that could remember numbers and could recognise simple arithmetic instructions it would be able to carry out complicated calculations automatically. He called the first calculating machine he invented the 'Difference Engine'. With this he could produce quite complicated tables of numbers. The machine did not have a memory. His next invention was never built. It was named the 'Analytic Engine'. Sadly for Babbage the engineers of his time were not able to produce a working model of his design, which was full of moving parts.
>
> What would Babbage have thought of one of the first computers to work? It was made in the USA in 1946. It filled a room and had about 18 000 radio valves. What could this fantastic machine do? It could store 20 numbers in its memory. Babbage would have been impressed. Since then the valves have been replaced by integrated circuits. These are much smaller and more reliable. Computers are becoming ever more powerful. There are new developments almost daily. Today many people own computers that are thousands of times more powerful than the computer imagined by Charles Babbage.

# chapter 11
# Sounds like ...

# Introduction

**Outcomes**

At the end of this chapter you should be able to:
- State what sound is and what causes sound.
- Carry out some experiments to demonstrate sound.
- Describe a sound wave and draw examples of different waves.
- State what frequency is and give some examples of common frequencies.
- Describe how sound travels through the ear and how we hear.
- Describe the human voice box and how we make sounds.
- State how sound travels through solids and liquids.
- State the speed of sound.
- Discuss noise and its effects in the workplace.
- Describe how musical instruments make sound.

Your teacher may give you a copy of these Outcomes for your workbook.

In this section you have been looking at energy. This chapter deals with sound, which is a form of energy caused by the movement of particles.

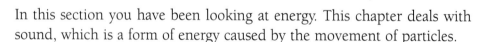

### Noise annoys

Your problem is to design a box that blocks out the noise from a source such as a small radio or a watch alarm.

1. Collect materials you think will absorb sound, such as newspaper, polystyrene foam, egg cartons and wood shavings, and a cardboard box.
2. Design and build your 'soundproof box'. Leave enough space for the radio, and don't forget to include a lid or door.
3. Work out a way of testing how well the insulation cuts down the noise.
4. Give your design a mark out of 10.
5. Write a report (with pictures) that describes your design, how well it did the job, how you measured the noise reduction, and any suggestions for improvement.

# What is sound?  11.1

The simple activities below will help you to understand some things about sound.

| NP | CSF |
|---|---|
| 3 | |
| 4 | |
| 5 | |
| 6 | |

### try this

**Making sounds**

1. Make the simple instruments as shown in the pictures.
2. Use the instruments to investigate how sound is produced.

**COLLECT**
- ruler
- G-clamp
- tuning fork
- block of wood
- fishing line or guitar string
- nails
- bottle
- polystyrene ball
- needle
- string
- balloon
- hand drill and drill bit
- yogurt container
- candle
- rubber band

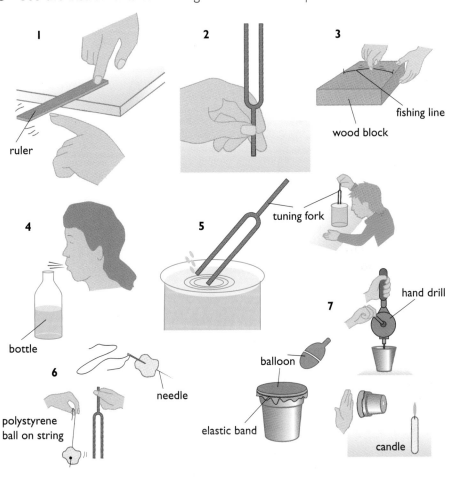

1. Describe sound and how it travels, using the word **'particles'**.
2. Explain what causes sound, using the word **'vibrations'**.
3. Explain why sound is referred to as **'sound energy'**.
4. For each activity that you did:
   - **a** state the source of the energy and the receiver
   - **b** describe how sound energy was transferred and whether sound energy was transformed.

### new words

particle • vibration • sound energy

Sounds like ...  139

# What does sound look like? 11.2

Sound is a form of energy that travels in waves. When a tuning fork is struck in the air, it causes vibrations. When the source of sound (the tuning fork) vibrates forwards, the vibration is passed to the molecules in the air. The air molecules are pushed together, so the air pressure is greater. This is called **compression**. When the tuning fork vibrates backwards, it leaves a space in which there are fewer molecules of air. The molecules spread out, so the air pressure is reduced. This is called **rarefaction**.

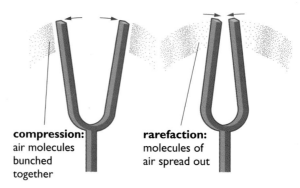

**compression:** air molecules bunched together

**rarefaction:** molecules of air spread out

This compression–rarefaction process repeats itself for as long as the vibration continues. This series of compressions and rarefactions form a **compression wave** or **longitudinal wave**. To imagine this, you can think of the coils in a 'slinky spring'.

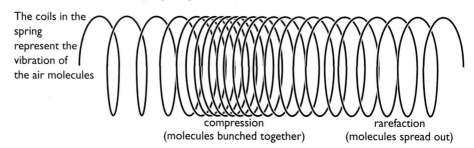

The coils in the spring represent the vibration of the air molecules

compression (molecules bunched together)

rarefaction (molecules spread out)

Sounds can be picked up by a **microphone** and fed into an **oscilloscope**. Oscilloscopes are instruments that allow us to see sounds as waves. An example is shown below.

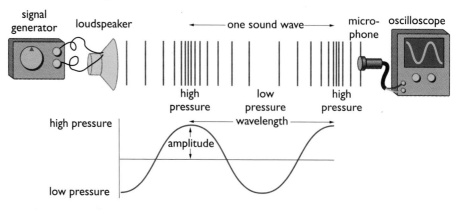

This type of wave is called a **transverse wave**.

140 ScienceMoves 2

### try this

**Looking at sound waves**

**COLLECT**
• oscilloscope

Different sounds produces different shaped waves.
- Low notes have long **wavelengths** (the distance from the peak of one wave to the peak of the next).
- High notes have short wavelengths.
- Loud notes have large **amplitudes** (the height of the wave).
- Soft notes have small amplitudes.

1. Draw diagrams to show the different sound waves you think would be produced by these sounds.
2. Use the oscilloscope to check your predictions.
3. Draw diagrams of **a** a soft low note **b** a high loud note.
4. Check these using the oscilloscope.
5. State the name of the wave shown on the oscilloscope. Explain whether sound waves really look like this.

### your turn

1. Fill in the missing words in the diagrams below.
2. Use the model of the 'slinky spring' to explain how sound travels.
3. Give an example of sound that includes a source of the energy (input), a receiver of the energy and the finishing energy (output), and show how the energy is transferred or transformed.

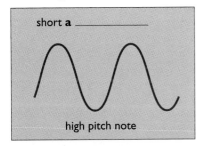
short a _____
high pitch note

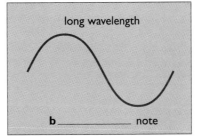
long wavelength
b _____ note

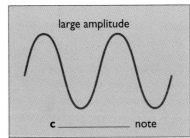
large amplitude
c _____ note

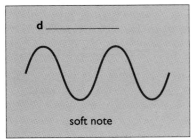
d _____
soft note

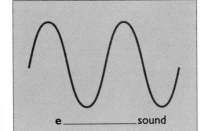
e _____ sound

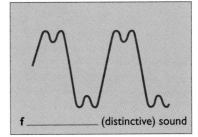
f _____ (distinctive) sound

### new words

**compression • rarefaction • compression wave • longitudinal wave • microphone • oscilloscope • transverse wave • wavelength • amplitude**

Sounds like ... 141

# The frequency of sound waves 11.3

By looking at sound waves, it is possible to measure how fast energy is transferred. This is done by measuring the number of **vibrations per second**, or **frequency**. Frequency is measured in **hertz** (Hz).

1 Hz = 1 vibration per second
1000 Hz = 1 **kilohertz** (kHz)

The frequency determines the **pitch** of a sound. For example, a high-pitched sound, such as an ambulance siren, has a high frequency. A low-pitched sound, such as a foghorn, has a low frequency.

Look at the diagrams below.

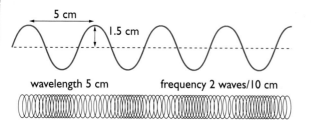
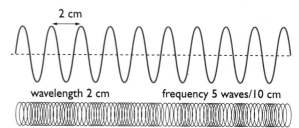

1  Describe what happens to the wavelength as the frequency increases.
2  Describe what happens to the wavelength as the frequency decreases.

## The range of human hearing

Different animals can make and hear different frequencies of sound. Humans can hear sounds ranging from 25 Hz (such as thunder) to around 20 000 Hz (20 kHz). Some animals are able to hear very high sounds. Bats and porpoises make sounds as high as 120 kHz and can hear sounds up to 150 kHz. A 'silent' dog whistle is usually in the range of 12–14 kHz, so if you listen carefully you should be able to hear one. Dogs can hear sounds up to 50 kHz, so they can hear many high-pitched sounds that are not audible to humans.

1  Design experiments to show that different animals have different hearing ranges.
2  Find the hearing ranges for three animals that are not mentioned here.

**new words**
vibration per second • frequency • hertz • kilohertz • pitch

# Sound and the ear     11.4

| NP | CSF |
|---|---|
| | 3 |
| | 4 |
| | 5 |
| | 6 |

1  Look at the diagram of the ear. Describe how sound travels through the ear. (You may wish to look back at the information about the ear on page 21.)
   Your teacher may give you some additional sheets to help you.

**OUTER EAR**
- **ear flap**—collects sound
- **wax**—prevents infection
- **ear drum**—vibrates and transmits sounds to the middle ear

**MIDDLE EAR**
- **ear bones**—amplify the sound and transmit it to the inner ear

**INNER EAR**
- **auditory nerve**—carries impulses to the brain
- **cochlea**—changes sound vibrations into nerve impulses (electrical messages)
- **Eustachian tube**—joins the middle ear to the back of the throat, and keeps the air pressure the same on both sides of the ear drum

2  Draw a flow chart to show how we hear sound.

## try this

### Keep your ears on the ball

Try to catch a bouncing ping-pong ball with and without earmuffs on.

**COLLECT**
- ping-pong ball
- earmuffs

1  Describe the difference in your ability to catch the ball. Try to explain how sound is involved.
2  Why do umpires ask the crowd to be quiet during tennis matches?

## project

# Sound wave

Use your library resources to research the following topic.
   Imagine you are a sound wave moving through the ear.
- What made you?
- What do you look like?
- How do you travel through the air?
- What is your frequency?
- How do you get into the ear?
- How do you travel through the ear?
- What organs can you see?
- How do you get to the brain?
- Will you cause damage to the eardrum?
- Are you recognised by the brain?

   Present this information in a creative way (for example, as a dance, mime or role-play).

Your teacher may give you an assessment criteria sheet.

# 11.4 continued...

### try this

**COLLECT**
- blindfold

### Direction of hearing

1. Copy the drawing below into your workbook.
2. Blindfold your partner. Do not speak. Snap your fingers or make a sharp noise at one of the points shown on your diagram.
3. Your partner describes exactly where the noise came from. Put a tick on the diagram for a correct description; put a cross for a wrong description.
4. Quietly move to a new position and repeat the test. Do this until most or all of the positions have been tested.
5. Now your partner can test your hearing in the same way.

1. When is it easy to judge where a sound has come from?
2. When is it difficult to judge where a sound has come from?
3. Describe what these experiments tell you about how the ear is involved in hearing and co-ordination.

### Human hearing and sounds

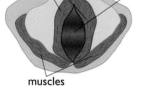

Not all sounds that reach the ear are heard. The **auditory nerve** transmits only one-twentieth of the sounds that reach the **cochlea**. So sounds are **filtered**.

Sounds travel from the ear to the brain. The brain compares the sound with stored memories so that we can make sense of the sound we hear. Adults can recognise about 500 000 sounds. If a new sound is heard, it is stored in the memory.

We make sounds by making our **vocal cords** vibrate at different frequencies. The vocal cords can change shape to alter the sound produced.

### try this

### Say 'ah'

1. Press two fingers lightly against the front of your throat and quietly say 'ah' several times. Shift your fingers until you feel the vibration of your vocal cords.
2. Change the pitch of the sound. Feel the change in your vocal cords.
3. Speak loudly and softly. What happens to the vocal cords?

1. Describe how the vocal cords change when you speak and make sounds.
2. Give a reason why males generally have lower voices than females.
3. Investigate the vocal cords and add any new words to your glossary.

**auditory nerve • cochlea • filter • vocal cords**

# Other features of sound    11.5

Sound can be transferred through materials other than air. Find out how well sound is transferred through solids and liquids by carrying out the following investigations.

| NP | CSF |
|---|---|
| 3 | |
| 4 | |
| 5 | |
| 6 | |

### try this

Predict what you expect to happen in each of these investigations.

### Ticking watch

1. Place a ticking watch about 30 cm from your ear. Listen carefully.
2. Place the wooden ruler against the watch and against your ear. Listen carefully.
3. Place the metal rod against the watch and your ear and again listen carefully.

### Wall intercom

Use a beaker to listen through a wall.

### String telephone

1. Use your cans and string and make a string-can telephone.
2. Talk to your partner through your 'telephone'.

### Watery sound

Investigate whether you can hear the sound of a bell in water. Can you hear it more clearly in water than in air?

### Vacuum radio

1. Turn the radio on and place it in the bell jar. Listen for the sound.
2. Connect the bell jar to a **vacuum** pump.
3. Start to pump the air out of the jar. Listen for the sound.

**COLLECT**
- ticking watch
- wooden ruler
- metal rod
- glass beaker
- two metal cans
- 2 m string
- glass tank
- small radio
- rubber bands
- bell jar
- hand bell
- bucket of water
- grease
- vacuum pump

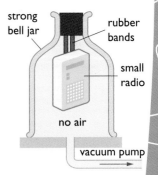

Write a report on each of the investigations. Include your aim, method and results. Were your predictions correct? Discuss your results using the following questions as a guide.

1. Did the wooden ruler and metal rod allow you to hear the ticking watch?
2. Explain whether sound was transferred better through the metal rod or the wooden ruler. Explain this in terms of the density of the materials. (**Hint:** Think how the particles are arranged in more dense objects and how this might affect sound waves.)
3. Could you hear your partner through the wall? Why?
4. What carried the sound waves in your string-can telephone? Why did you need a 2 m length of string?
5. Explain whether sound was transferred better in water or air. Give a reason for your answer.
6. Explain whether sound was transferred in a vacuum. Give a reason for your answer in terms of particles and vibrations.

# 11.5 continued...

## Speed of sound

Sound is transferred much more slowly than light. In air, the speed of sound is about 340 m/s. The **speed of sound** varies with temperature; sound is transferred faster in warm air. As you have seen with your earlier investigations, sound is transferred at different speeds in different substances, or **media**. It travels faster in more dense media. Thus, sound travels faster in solids and liquids than in gases, because the particles are closer together.

The table opposite shows the speeds of sound in different media.

| Medium | Speed of sound (m/s) |
|---|---|
| Air (room temp.) | 343 |
| Water | 1480 |
| Wood | 3400 |
| Polythene | 540 |
| Glass | 4900 |
| Iron | 2800 |
| Hydrogen | 1310 |
| Steel | 5020 |

### try this

**COLLECT**
- stopwatch
- measuring tape

### Timing sound

1. Stand exactly 50 metres from a wall that gives a good echo.
2. Make a loud clap and listen for the echo. (It takes less than a second for the sound to get to the wall and back – too short a time to measure with a stopwatch.)

3. Make a second clap the moment you hear the echo. Keep this up, clapping with a rhythm that makes each clap cover the echo of the previous clap.
4. Time 20 of these claps, counting the first as '0'. In this time the sound will have travelled 2000 metres (20 journeys of 100 metres). From this you can calculate its speed using the formula:

$$\text{speed} = \frac{\text{distance}}{\text{time}}$$

### your turn

1. Why is sound transferred faster in water than in air?
2. Using the speed of sound as a guide, list the following in order of least dense to most dense: hydrogen, steel, polythene, wood, air, water, glass.
3. Explain why, in outer space, communication is by radio waves, not sound waves.
4. Why can you see lightning before you hear the thunder?
5. The speed of sound in air at sea level is about 300 m/s at 0°C. If the ABC used sound waves instead of radio waves for their broadcasts, how long would it take a program to reach Melbourne from Sydney (885 km)? What other problems would there be?

### new words

vacuum • speed of sound • media

# What makes a note? 11.6

Musical instruments rely on vibrating air to make sound. Different shapes and sizes of instruments produce different sounds.

**try this**

### Straw pipe

1. Trim the drinking straw to make a mouthpiece as shown in the diagram. Blow through the straw. Shorten the pipe by cutting the end off the straw.
2. State what happens to the pitch (frequency) of the note when you shorten the straw.

### Test tuba

1. Set up a row of test tubes containing different levels of water.
2. Hold each test tube in front of your lip and blow into it.
3. Describe what causes the sound in the test tubes.

### Hose notes

1. Blow across a piece of rubber hose in the same way.
2. Repeat with narrower and wider pieces of hose.
3. Explain how the length, diameter, type of material and thickness alter the pitch of the sound.

### Sound board

1. Make a sound board as shown in the diagram.
2. Experiment by moving one of the wedges and listening to the sound made by the string.
3. Describe what happens to the note and how it changes.
4. Describe what happens to the string as it is tightened.

1. Make a list of 20 musical instruments. State which of the methods you have investigated is used to produce sound in each of the instruments in your list.
2. Make your own musical instrument that can produce different sounds. How does your instrument make low- and high-pitched sounds? Demonstrate this.

**COLLECT**
- scissors (sharp enough to cut through straws)
- drinking straws
- test tubes and rack
- rubber hose of different diameters
- fishing line
- wooden board approx. 30 cm × 8 cm
- wooden wedges

| NP | CSF |
|---|---|
| 4 | |
| 5 | |
| 6 | |
| 7 | |

Sounds like ... 147

## 11.6 continued...

### Pitch, tone, timbre and resonance

As we have seen, the frequency of a sound determines its pitch. In music, we express the pitch of a note as being 'high' or 'low'. Sounds can also vary in **intensity**; that is, whether they are loud or soft.

The quality of a sound or note is called its **tone**. Some musical instruments have a bright, light tone, while others have a heavy or more subdued tone.

Different instruments can play the same note at the same intensity, but still sound different. The distinctive sound of an instrument is called its **timbre**.

Many objects continue to vibrate at their natural frequency after they have been struck or played. This continuing vibration is called **resonance**. For example, when you 'ping' a glass it vibrates at its natural frequency. An opera singer may be able to produce a sound of the same frequency, which makes the glass vibrate so much that it shatters.

### try this

**COLLECT**
- signal generator
- loudspeaker
- light plastic balls

**Resonating balls**

1. Connect the loudspeaker to the signal generator as indicated.
2. Put the plastic balls into the loudspeaker cone. When the loudspeaker is vibrating at its natural frequency, the balls will jump about.

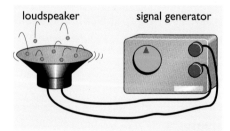

**new words**

intensity • tone • timbre • resonance

# Turn that noise down    11.7

We are bombarded by all sorts of sounds every day. Sounds that are nice to some people may be unpleasant to others.

### your turn

1. List 10 pleasant sounds and 10 unpleasant sounds. Compare your list with those of other students. Ask your family members to make their own lists, and compare your list with theirs also.
2. Explain whether all people consider the same sounds to be pleasant and unpleasant.

The loudness of a sound depends not only on the sound, but also on the person who hears it. Because of this, scientists don't measure loudness directly; instead, they measure **sound intensity**. This is measured in **decibels** (dB) using a **sound level meter**. The decibel levels of some everyday sounds are shown in the table. An increase of 10 dB means a sound will seem twice as loud. For example, a vacuum cleaner (80 dB) sounds twice as loud as heavy traffic (70 dB).

| Sound | Decibels |
|---|---|
| Something you can just hear | 0 |
| Your own breathing | 10 |
| Leaves rustling in the breeze | 20 |
| A suburban street at night | 40 |
| Two people talking | 60 |
| Busy traffic | 70 |
| A vacuum cleaner | 80 |
| An electric drill 20 cm away | 90 |
| An underground train | 100 |
| A rock band | 90–140 |
| A machine gun 3 metres away | 130 |
| A jet plane taking off overhead | 140–150 |

Exposure to sounds of 90 dB for a long period of time can damage your ears. Exposure to sounds of 110 dB for more than 2 minutes can result in permanent deafness.

### your turn

1. Draw a graph showing the decibel levels of the sounds shown in the table above.
2. Describe how sound level is measured.
3. Describe how the decibel level is related to the loudness of sound.
4. Use the table to estimate the sound level in your classroom.
5. Explain why earmuffs are worn by many people in machinery workshops.

HEARING PROTECTION MUST BE WORN

Your teacher may give you a handout on sound level meters.

## Hearing defects

There are many types of deafness. Some people are born deaf (**congenital deafness**). Others become fully or partly deaf through accidents, infections or exposure to excessively loud noise. This is known as **hearing loss**. There are two basic types of hearing loss.

### Outer and middle ear

The ear canal can become filled with wax or an infection can cause a build-up of fluid. Both of these conditions block the passage of sound vibrations to the eardrum, so hearing is impaired.

The eardrum or middle ear bones can be damaged by very loud noises, such as explosions, or by a knock on the head. Damage to the outer and middle ear can be treated or the hearing can be assisted by the use of a hearing aid.

### Inner ear

Inside the cochlea are tiny hair-like fibres that respond to the vibrations and act as the receptor cells. These hairs can be damaged, usually by exposure to excessively loud noise, such as jet planes, racing cars, rock bands or noisy work environments. The damage and destruction of these fibres are irreversible and cannot be overcome with the use of a hearing aid.

### on your own

**Can you hear me?**

1. Imagine you had a brother or sister who was deaf. Design an experiment to demonstrate to them that sound is carried by vibrations. They must be able to feel the vibrations when the sound is made.
2. Investigate how people whose hearing or speech is impaired are taught to communicate. You could find out about the different types of sign language, and then try using them yourself.

 Your teacher may give you a handout on hearing defects.

**new words**

sound intensity • decibels • sound level meter • congenital deafness • hearing loss • audio oscillator

# Hearing loss

## 11.8

Read the following article from *New Scientist* magazine.

### France takes steps to turn down the volume

The intolerable tinny strains of someone else's personal stereo may become a thing of the past, in France at least. Last week the National Assembly voted to put an end to ear-splitting stereos and limit their output to 100 decibels.

Deputies say the move is designed not so much to improve life for other Metro passengers as to prevent hearing loss among young people. 'We're producing a generation of deaf people,' deputy Jean-Francois Mattei told colleagues.

The amendment has still to be approved by the Senate. If it becomes law, all personal stereos sold in France will also have to carry a warning that 'prolonged use at full power can damage the ear of the user'.

Jean-Pierre Cave, who is an ear surgeon as well as a deputy, says that listening to a personal stereo at volumes above 100 decibels can lead to irreversible damage to the ear after a few hours. Above 115 decibels, damage can occur within minutes.

FNAC, a leading electronics retailer, says most of the personal stereos it sells have an output of more than 100 decibels, and some can produce 126 decibels. But the company doubts the amendment could become law because there are no accepted international standards against which to check equipment.

The deputies decided to take action after several studies suggested an up-surge in the number of young people suffering some degree of deafness. One recent survey of 400 youths found that 1 in 5 had some hearing loss. A similar study a decade earlier found only 1 in 10 suffered some loss.

Another French study by hearing specialist Christian Meyer-Bisch found that rock concerts are more likely to damage hearing than listening to personal stereos or even going to clubs (*This Week*, 27 January, p. 5). He called on the government to limit output at concerts to 100 decibels.

Cave says the proposed legislation on personal stereos is the 'first step'. He is busy preparing amendments that would also turn down the volume at clubs and concerts.

Tara Patel, Paris

| NP | CSF |
|---|---|
| | 4 |
| | 5 |
| | 6 |
| | 7 |

1. Explain why the French want to put an end to 'ear-splitting stereos'.
2. Explain why France is 'producing a generation of deaf people'.
3. List the scientific words you are familiar with in this article.
4. State the scientific evidence mentioned in the article that has prompted the action.
5. Explain how science can influence the laws in a society.
6. Explain whether you think similar changes should be made to the law in Australia.

Your teacher may give you a table on hearing protectors.

## new words

Add any new words to your glossary.

# another look 11.9

## 11.1 What is sound?
1 Explain what causes sound, giving examples from the experiments you carried out.

## 11.2 What does sound look like?
1 Describe what sound 'looks like'.

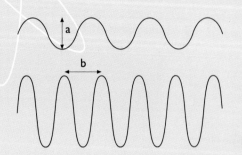

2 Draw a diagram to illustrate compression and rarefaction when a tuning fork is struck.
3 Look at the diagram opposite.
   a Label a and b.
   b Of the two sound waves, which has the:
     • highest frequency?
     • lowest amplitude?
     • longest wavelength?
   c Of the two sound waves, which would produce the:
     • lower pitch sound?
     • loudest sound?

## 11.3 The frequency of sound waves
1 Define frequency and state the unit used to measure it.
2 Explain how wavelength is affected by frequency.
3 State the range of human hearing and explain whether animals can hear better than humans.

## 11.4 Sound and the ear

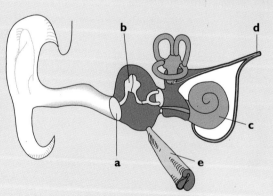

1 a Label the parts a–e in the diagram opposite.
  b Draw a flow chart of how the ear hears and recognises sounds.
2 Explain how co-ordination is affected by hearing.
3 Explain:
   a whether humans hear all sounds
   b how they make sounds.

## 11.5 Other features of sound
1 Briefly describe how sound travels through different media. Can it travel in a vacuum?
2 State the speed of sound in air.

## 11.6 What makes a note?
1 Explain how sounds are produced in a trumpet, a flute and a guitar.
2 Give two examples of resonance.

152 ScienceMoves 2

## 11.7 Turn that noise down

| Decibels (dB) | Noise source |
|---|---|
| 0 | threshold of hearing |
| 10 | leaves rustling |
| 20 | inside a broadcasting studio |
| 30 | soft whisper |
| 40 | residential area at night |
| 50 | |
| 60 | inside a normal home and office |
| 70 | normal conversation |
| 80 | cars |
| 90 | trucks, a printing press plant |
| 100 | inside a foundry or boiler room |
| 110 | pneumatic jackhammer |
| 120 | jet plane take off (at 70 m) |
| 130 | threshold of painful hearing |
| 140 | Mirage jet fighter (at 2 m) |
| 180 | eardrum ruptures |

1. Look at the table above.
   a. How much louder is a jackhammer than a conversation?
   b. How much louder are heavy trucks than cars?
   c. If you were exposed to boiler room noise all day, what would be the effect on your hearing?
2. Explain who would hear sounds first in the following situations.
   a. A bell is rung underwater:
      - a person swimming 3 m from the bell
      - a person in the air 3 m above the bell.
   b. a steel girder on a tall building is tapped:
      - a person standing at the bottom of the building
      - a person, also at the bottom of the building but with their ear against the steel girder.
3. Explain why lightning is seen before thunder is heard.

Your teacher may give you a word puzzle to complete.

## Linking concepts

Look back at the work you have done in chapters 9, 10 and 11. Draw a concept map to show how these chapters link together.

## Strengths and weaknesses

Don't forget to identify your strengths and weaknesses in this section of work.

Your teacher may give you a summary sheet and a chapter test.

# Mathematical calculations

The speed of sound can be calculated using the formula:

$$\text{speed of sound (m/s)} = \frac{\text{distance (m)}}{\text{time (s)}}$$

Echoes are often used to help calculate the speed of sound in different media. When using echoes, the distance has to be doubled because it represents the time taken to get to the wall or cliff, and the time taken to travel back.

Use the formula to solve these problems. Assume sound travels through air at 330 m/s.

1. A boy and a girl are standing at the foot of a steel-framed building under construction. A worker 66 m up in the building is slowly hammering a new steel girder into position. The girl hears the sound through the air.
   a. How long does the sound take to reach the girl?
   b. The boy has his ear to the steel frame. If sound travels through steel at 4950 m/s, how long would it take the sound to reach the boy?

2. A ship is using sonar to find the depth of the sea water. It sends a sound signal down to the sea bed where it is reflected and travels back to the hull of the ship. A microphone detects the sound three-fifths of a second after the signal was sent. How deep is the water? (Assume sound travels in sea water at 1500 m/s.)

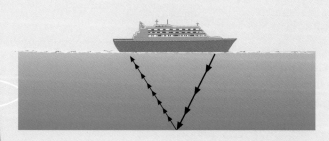

3. During a thunderstorm, a girl noticed that she heard thunder exactly 5 seconds after she saw a lightning flash. How far away was the electrical storm that caused both the thunder and the lightning?

4. A stone is dropped down a well. The splash is heard 0.30 seconds after dropping the stone. How deep is the well?

5. Hearing-impaired runners use the smoke of the starting pistol rather than the sound when competing in races. If the starter stands 20 m away, how much quicker off the blocks is the hearing-impaired runner than the other runners?

Earth and beyond

section five
# Rocks, earth and sky

science in context
## 12 Sun, moon and sky

## 13 Look beneath your feet

# Chapter 12

# Sun, moon and sky

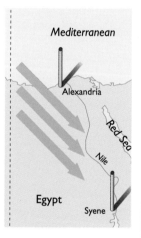

If the Earth were flat, the stick would cast the same length shadow in Alexandria and Syene.

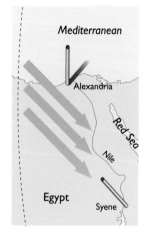

On a curved Earth, the Sun casts a shadow in Alexandria but not in Syene, where it is directly overhead.

# Introduction

### Outcomes

At the end of this chapter you should be able to:
- Describe the Earth, Sun and Moon.
- State why we have day and night and seasons.
- Describe and draw a solar eclipse and a lunar eclipse.
- Discuss events caused by the movement of the Sun, Moon and Earth.
- Explain time zones.
- State how low and high tides occur.
- Explain what early astronomers believed about the Sun, Moon and Earth.
- Describe an Aboriginal explanation of the relationship between the Sun, Moon and Earth.

Your teacher may give you a copy of these Outcomes for your workbook.

Throughout the ages people have tried to explain how the Sun rises, why the Moon shines and why the Earth is the way it is. In this section you will look at early explanations of the world and what we understand today. You will then explore the Earth in more detail.

### Ideas left behind

In early times, the Earth was thought to be flat. In ancient Greece, Aristotle (384–322 BC) first suggested that it was a sphere. He noticed that the position of the stars shifted as he travelled from north to south, and that during a lunar eclipse the Earth cast a curved shadow on the Moon.

A fellow Greek, Eratosthenes (276–194 BC), showed by an experiment that the Earth cannot be flat. He argued that if the Earth were flat and the Sun a long way off, identical vertical sticks would cast shadows of the same length anywhere in the world. He measured such shadows at midday on the same day of the year in two Egyptian cities 800 km apart. They had different lengths, so the Earth could not be flat.

156  ScienceMoves 2

# Aboriginal beliefs                                    12.1

Each group of people tries to explain the world around them. The following is an adaptation of an Aboriginal creation story.

| NP | CSF |
|---|---|
| 3 | |
| 4 | |
| 5 | |
| 6 | |

> Yhi the Sun god waited for Baiame the Great Spirit to tell her to go down to Earth. Yhi floated down and as she did her light spread into all the places hidden by darkness. Flowers, trees and shrubs sprang to life, insects opened their wings and flew about in her light. Ice melted and lakes overflowed, giving water to thirsty plants; fished played in the water and reptiles and snakes found homes on dry land.
>
> All animals – those of fur and feathers – came out of their caves and danced in the light. Yhi told the animals that she would send them summer which would ripen fruit and winter for sleeping through cold winds. Then she rose and left the Earth and sank below the western hills. All the creatures were very sad.
>
> Hours later there was a twittering of birds announcing that Yhi had returned to flood the Earth with light once more. Yhi realised the creatures were afraid in the darkness, so she sent the morning star to tell the world she was coming and during the night she gave them Bahloo, the Moon. The creatures on Earth were happy when the Moon sailed through the night sky, giving birth to stars and making a wondrous glory in the heavens.

Adapted from *Aboriginal Stories of Australia* by A. W. Reed (Angus & Robertson, 1993)

1. Explain how this group of Aborigines explained the rising and setting of the Sun.
2. Describe what the Sun brought to the Earth.
3. State what season it is when most flowers start to grow and lakes overflow with water.
4. Describe what Yhi promises she will bring.
5. Explain how this story describes the Moon and the stars.
6. Explain why you think this story was told.
7. List any questions you would like to ask about this story and the account on the previous page to help you understand them better.

**new words**

Add any new words to your glossary.

# Movement of the Earth and Moon   12.2

| NP | CSF |
|---|---|
| 3 | |
| 4 | |
| 5 | |
| 6 | |

### Day, night and the seasons

The Earth spins from west to east as it **orbits** the Sun. It spins around an imaginary 'axle', or **axis**. One **rotation** takes 24 hours. Because of this rotation, only one side of the Earth is facing the Sun at any one time. The other side experiences darkness, or night.

The Earth's axis is tilted 23.5° in relation to the Sun. This tilt means that one part of the Earth is closer to the Sun than the rest, and so receives more sunlight. As the Earth **revolves** around the Sun, the part tilted towards the Sun experiences summer and the part tilted away experiences winter.

Refer to the following diagrams to answer these questions.
1. Describe the alignment of the Earth on its axis.
2. Explain what causes day and night.
3. Explain what causes the seasons.

Your teacher may give you additional diagrams to help you.

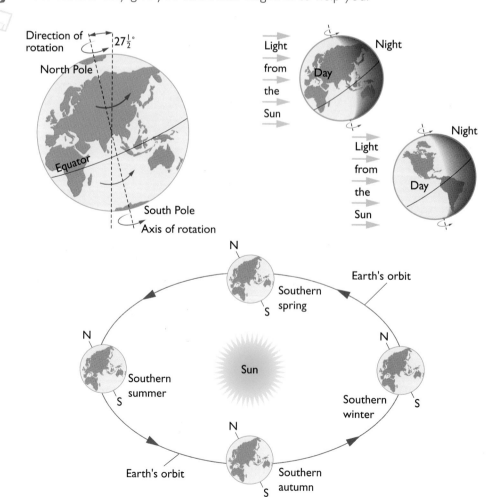

158  ScienceMoves 2

1. Explain why the Sun rises in the east and sets in the west. You may like to use a model of the Earth and a torch to show this.
2. Days are of different lengths in different places on Earth. One factor that determines this is distance from the Equator (an imaginary line around the middle of the Earth). State two other factors that have been mentioned.
3. The Earth takes 365¹/₄ days to revolve around the Sun. Describe what happens to the extra quarter each year. (**Hint:** Think of leap years.)

## The Moon

The Moon's gravity causes the Earth to 'wobble' slightly in its orbit around the Sun. The Moon orbits around the Earth and the two together orbit the Sun.

The Moon takes about 29 days to go around the Earth. It takes the same time to spin on its axis. From Earth we can only ever see the same side of the Moon. The different shapes the Moon takes in the sky are called the **phases** of the Moon. The Moon has no light of its own, but reflects the Sun's light back to Earth.

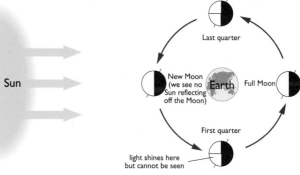

The phases of the Moon.

Copy the diagram of the phases of the Moon as it revolves around the Earth.
1. Draw the shape of the Moon as it would be seen from Earth in each position.
2. Label your drawings **full moon**, **new moon**, **first quarter** and **last quarter**.
3. In between the four phases you have drawn, there are four other phases, or shapes, of the Moon. These are known as **crescent** and **gibbous** phases. Add these to your drawing and label them.

**on your own**

## Moon research (1)

The rising and setting times of the Sun and Moon are printed in the daily newspapers. Collect these times over a week. Draw a picture of the Moon for each day, and state what phase the Moon is in.

Your teacher may give you a sheet to record this information.

## Step back in time

In the 1960s, space programs were already under way to send people to the Moon. The Saturn V, America's most powerful rocket, was used to achieve this task. The first Saturn V lifted off on 9 November 1967; it was unmanned. The first manned flight was *Apollo 8*, launched in December 1968, which orbited the Moon but did not land. *Apollo 11*, launched on 16 July 1969, carried Astronauts Neil Armstrong, Edwin ('Buzz') Aldrin and Michael Collins. After three days they reached the Moon and went

into orbit around it. On 20 July, Armstrong and Aldrin left the 'command module' of their spacecraft in a 'lunar module' called Eagle to land on the surface of the Moon. The famous saying, 'the Eagle has landed' was radioed by Armstrong later that day.

All around the world, people sat glued to their television screens watching images of the historic events. Six hours after landing, Armstrong set foot on the Moon saying, 'That's one small step for man, one giant leap for mankind'. Together, Armstrong and Aldrin unveiled a plaque that said: 'Here men from the planet Earth first set foot upon the Moon, July 1969 AD. We came in peace for all mankind.'

**on your own**

## Moon research (2)

Choose one of the following activities.

1. Conduct an interview with a person who can remember the 1969 Moon landing. Record what they can remember, and how they felt. Present your findings to the class.
2. Write your own story about the Moon landing, based on information from the time and any additional information you think important. You could write it from the point of view of one of the astronauts, or of someone waiting back on Earth for news.
3. Research the history of human exploration of the Moon. What prompted the space programs? How many Moon landings have there been? What discoveries were made about the Moon? Prepare a written report.

**new words**

orbit • axis • rotation • revolve • phase • full moon • new moon • last quarter • first quarter • gibbous • crescent

# The heavens

## 12.3

The movement of the Sun and Moon have other effects on the Earth and produce some interesting sights.

### Tides

In the 1600s, Isaac Newton put forward a theory about **gravity**: big objects exert a force on smaller objects, and closer objects exert more force than distant objects. Both the Sun and the Moon exert a pulling force on the Earth. The force of the Moon is greater because it is much closer to Earth. This force does not have much effect on the solid parts of the Earth, but it does pull the water in the oceans (and the gases in the atmosphere) towards the Moon, creating a 'bulge'.

In one day the Moon stays in roughly the same place so the bulge stays in roughly the same place. The Earth rotates on its axis once a day, passing through two high tide bulges and two low tides.

High tide and low tide (above) are caused by the gravitational pull of the Moon and the Sun.

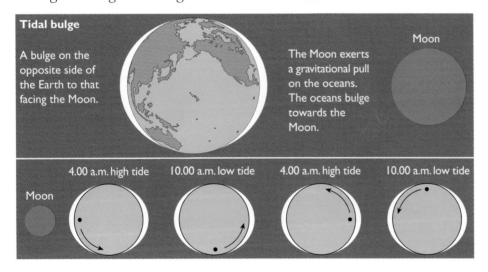

### Shadows in space

When you stand in sunlight your body forms a shadow on the ground. In the same way, the Earth and the Moon can also cast shadows on each other when they are in line with the Sun. These giant shadows are called **eclipses**.

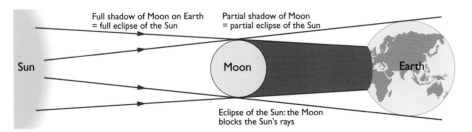

When the Moon passes directly between the Sun and the Earth, the Moon blocks out the light of the Sun, casting a shadow on the Earth. This is called an eclipse of the Sun, or a **solar eclipse**.

Solar eclipses occur when the Earth, Moon and Sun are all in line.

**Sun, moon and sky**

# 12.3 continued...

When the Earth moves directly between the Sun and the Moon, the Earth casts a shadow on the Moon. This is called an eclipse of the Moon, or a **lunar eclipse**.

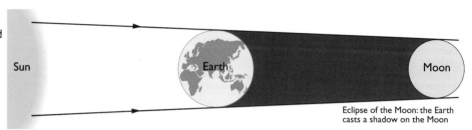

Eclipse of the Moon: the Earth casts a shadow on the Moon

## try this

**COLLECT**
- light bulb
- battery
- 2 connecting wires
- 2 balls

### Investigating eclipses

1. Make a model to investigate the various ways that the Sun, Moon and Earth could line up. Use the:
   - bulb to represent the Sun
   - larger ball to represent the Earth
   - smaller ball to represent the Moon.
2. Hold the 'Earth' in one hand. As you walk around the 'Sun', move the 'Moon' around the 'Earth'.

1. Which positions would cause the highest and lowest tides?
2. Which positions cause **a** an eclipse of the Moon? **b** an eclipse of the Sun?

## your turn

1. Describe three effects of the Sun and Moon that can be seen from Earth.
2. Draw the positions of the Sun, Moon and Earth that would produce:
   - the highest and lowest tides
   - an eclipse of the Sun
   - an eclipse of the Moon.
3. Explain why high tides are not exactly 12 hours apart.
4. Explain why an eclipse of the Moon or an eclipse of the Sun does not occur every month.
5. Explain how the smaller Moon can appear to block out the Sun and cause a solar eclipse.
6. Using the diagrams on this page and the previous page, describe solar and lunar eclipses.
7. Use the library to find out when the last solar and lunar eclipses were in your area.

## new words

gravity • eclipse • solar eclipse • lunar eclipse • partial eclipse

# Time zones

## 12.4

Our measurement of time depends on the rotation of the Earth. The Earth rotates every 24 hours from west to east, so the Sun appears to move from east to west. Because of this, places in the east see the Sun before places in the west. Places in the east are therefore 'ahead' in time compared to places in the west.

| NP | CSF |
|----|-----|
|    | 3   |
|    | 4   |
|    | 5   |
|    | 6   |

### on your own

Find out more about time zones. How many time zones is the world divided into? Which country spans the most time zones? Where is the time measured from? What is the 'date line'?

Your teacher may give you worksheets on time and/or an assessment criteria sheet for the project below.

# Time problems

## project

Choose one of the following problems to solve.

1. A sundial is an ancient and efficient way of telling the time. Find out about the sundial, when it was invented, who used it and any other details you consider important.

    Your teacher may give you an instruction sheet for making a sundial.

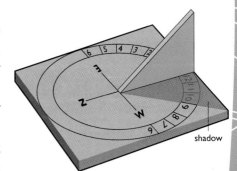

2. Justin and Sue wondered how people measured short time intervals before clocks were invented. Can you build an instrument that could accurately measure a minute.

    (**Hint:** Think of egg timers.)

3. Minh noticed that her shadow was longer than her body at certain times of the day. Can you help Minh find out when her shadow is exactly the same length as her body?

    (**Hint:** Use a metre ruler in the Sun and watch carefully.)

4. Dylan and Jackson were camping. Jackson was trying to explain to Dylan how the different shapes of the Moon occurred. He said it was because the Moon travelled around the Earth and reflected the Sun's light.

    Can you design a model to help Jackson explain this to Dylan.

### new words

**Add any new words to your glossary.**

# science in action

# The space shuttle
## 12.5

| NP | CSF |
|----|-----|
| 3  |     |
| 4  |     |
| 5  |     |
| 6  |     |

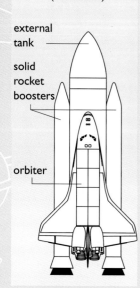

**Space shuttle (56 metres)**

- external tank
- solid rocket boosters
- orbiter

When men travelled to the Moon for the Apollo missions, rockets were used. Today space shuttles carry people into space. A space shuttle is made of a reusable space plane called an orbiter, two rockets and an external fuel tank. The fuel in the external tank and the rockets is burned to produce 3000 tonnes of thrust for lift-off.

After two minutes the shuttle has reached an altitude of 48 kilometres and the rockets have burnt out. They detach from the orbiter, parachute back to Earth and are collected for reuse. When the orbiter reaches its correct orbit, the external tank is jettisoned, and burns up in the atmosphere. The orbiter continues to orbit the Earth for the duration of the mission (usually four to seven days).

When rockets were used for the Apollo missions, the re-entry into the Earth's atmosphere was hazardous. The module carrying the astronauts fell through the atmosphere like a meteorite, protected from burning up by heat shields. It was slowed down by parachutes and then landed in the ocean, where the astronauts were then collected by boat.

The space shuttle lands on one of the world's longest runways at the Kennedy Space Center. The shuttle orbiter relies on a high-speed glide to bring it in to land. It must be a perfect landing because, unlike aeroplanes, it cannot circle and try again. All the propulsion equipment has been lost. The runway is covered with grooves to provide a more skid-resistant surface and is slightly sloped to get rid of water. The orbiter is protected from heat on re-entry by tiles, gap fillers and insulation blankets. This insulation also protects the orbiter and its occupants from the cold in space.

### your turn

1 Explain why objects burn up on re-entry to the Earth's atmosphere.
2 Explain how technology has changed space exploration and the way we view the world.
3 Explain whether you think the first people in space were brave people.
4 Explain how science improves our knowledge of the world and improves the technology we use in our lives.

### new words

Add any new words to your glossary.

# another look

## 12.6

### 12.1 Aboriginal beliefs

1. Explain why the ancient Greeks thought the Earth was flat.
2. Explain what caused them to believe that the Earth was round.
3. Describe how one group of Aborigines explained the rising and setting of the Sun, the seasons and the Moon.
4. Explain why people try to explain the world around them.

### 12.2 Movement of the Earth and Moon

1. Explain what causes day and night.
2. Explain what causes the seasons.
3. Explain what causes the phases of the Moon.
4. Describe a gibbous Moon.
5. Who were the first people to walk on the Moon? What spacecraft did they use? Describe their mission.
6. Explain why we have a leap year every four years.
7. State how long it takes:
   a. the Moon to orbit the Earth
   b. the Earth to orbit the Sun
   c. the Earth to turn on its axis.
8. Explain why people say they can see a face in the Moon.

### 12.3 The heavens

1. Explain what causes the tides. Draw a diagram showing how tides occur.
2. Describe the difference between a solar eclipse and a lunar eclipse.
3. The diagram below shows a full solar eclipse and a partial solar eclipse.
   a. Describe the difference between the two.
   b. Describe what you would see on Earth.

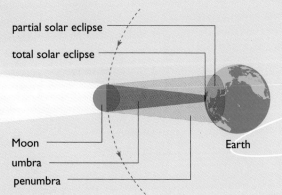

### 12.4 Time zones

1. Describe time zones and state how many zones the world is divided into.
2. Describe the date line.

3 Explain what happens to time when you travel:
   a from east to west
   b from west to east.
4 Explain why all places don't experience the same time.
5 Describe what problem you chose to do and how you solved it.

## 12.5 Science in Action: The space shuttle

1 State how the space shuttle is different to Saturn V.
2 Explain how technology has changed space flight and our lives.

## Movement of the Moon

1 Copy the following diagrams into your workbook.
2 Label the diagrams to show which would produce an eclipse of the Sun and which would produce an eclipse of the Moon.
3 Label the diagrams to show the position of a new moon and the position of a full moon.

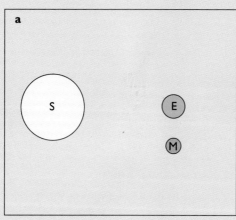
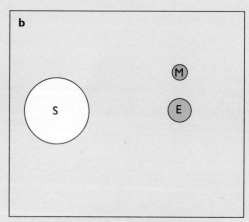
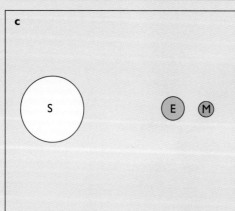
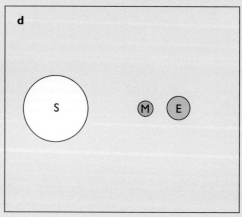

Your teacher may give you a summary sheet and a chapter test.

47, 48

### Strengths and weaknesses

Take time to review your strengths and weaknesses.

# Skillbuilder
# Using tables and data

An important skill in science is to be able to put data into a table. Tables allow you to compare information, both within a single table and between different tables, and to see relationships.

Data can come from a variety of sources. Tide charts are used by anglers to help them find the best times to go fishing. These charts state the low and high tide times at particular places. The tide times for a 24-hour period are found in the weather section of most daily newspapers.

Look up the tide times in your daily newspaper for the next two weeks. Create a tide chart by putting the tide times into a table of high and low tides.

### The tides

TODAY
Williamstown – high water: 5.10 am, 6.42 pm. low water: 12.11 am, 11.58 pm.
　Port Phillip Heads – high water: 2.21 am, 3.32 pm. low water: 8.30 am, 9.00 pm.
　Tooradin – high water: 3.44 am, 4.55 pm. low water: 9.53 am, 10.23 pm.

TOMORROW
Williamstown – high water: 5.54 am, 7.17 pm. low water: 12.48 am, 12.34 pm.
　Port Phillip Heads – high water: 3.01 am, 4.08 pm. low water: 9.12 am, 9.38 pm.
　Tooradin – high water: 4.24 am, 5.31 pm. low water: 10.35 am, 11.01 pm.

| Date | High tide | Low tide | High tide | Low tide | High tide | Low tide |
|---|---|---|---|---|---|---|
| | | | | | | |
| | | | | | | |

1. Work out the average time between each high tide over the two-week period.
2. Explain why this is not exactly 12 hours.
3. Can you find a pattern for the tide times?

In a separate table, record the Moon rising and setting times for the same two-week period.

| Date | Moon rise | Moon set |
|---|---|---|
| | | |

4. Find the relationship between the high and low tides and the Moon rising and setting times.
5. Explain which tide occurs when the Moon is directly overhead.

chapter 13

# Look beneath your feet

## Introduction

> **Outcomes**
>
> At the end of this chapter you should be able to:
> - Describe the three types of rocks found in the Earth.
> - Describe how the different rocks on Earth were formed.
> - Describe some physical processes.
> - State some common uses for the different rocks studied.
> - Explain what minerals are and how they can be identified.
> - Identify different types of rocks.
> - Make your own key to classify different types of rocks.
> - Make some rocks.
>
> Your teacher may give you a copy of these Outcomes for your workbook.

In the last chapter you looked at the Sun and Moon and their relationship to the Earth. In this last chapter you are going to look more closely at the different types of rocks that make up the Earth.

Volcanoes erupt, spewing out lava which solidifies to produce new rock formations.

Glaciers erode valleys and transport material.

# Three types of rocks  13.1

Scientists have learnt a lot about the Earth by studying rocks. They have found out about life on Earth and the changes the Earth has gone through. There are three different types of rocks: sedimentary, igneous and metamorphic rocks.

| NP | CSF |
|---|---|
| 4 | |
| 5 | |
| 6 | |
| 7 | |

### try this

## Exploring rocks

**COLLECT**
- igneous rocks (scoria, granite)
- sedimentary rocks (sandstone, limestone)
- metamorphic rocks (marble, hornfels)

1. Look at the samples of scoria, sandstone and marble. Describe these rocks. What are the differences between them?
2. By examining each rock, state how you think it was formed.
3. Now look at the samples of limestone, granite and hornfels.
   a State what sedimentary rocks have in common.
   b State what igneous rocks have in common.
   c State what metamorphic rocks have in common.
4. Using the rocks you have here, make a simple key to help somebody identify the different types of rock. Use the following criteria: small grains or large grains; crystals or no crystals; large crystals or small crystals; etc.
   (If necessary, look back at *ScienceMoves 1* to help you make a key.)

Your teacher may give you a rock key to help you.

Igneous rock (granite).

Sedimentary rock (sandstone).

Metamorphic rock (hornfels).

## Sedimentary rocks

As rocks age and are exposed to the weather, they fall apart and crumble, forming particles that can be washed away into rivers and waterways. These particles can be carried by the river until they eventually settle and form a **sediment** on the bottom of the river. Over millions of years this sediment is squashed, heated and subjected to chemical action. It undergoes tremendous pressure and hardens to become a rock.

**Sedimentary rocks** are easy to recognise because they are made of bits and pieces. Sedimentary rocks can be made:
- mechanically – by wind and running water
- chemically – by chemical changes in the sediment (as with limestone, which is made of calcium carbonate crystals), and
- organically – from living things such as coal, peat and coral.

Your teacher may give you a handout on sedimentary rocks.

Look beneath your feet  169

## Igneous rocks

'Igneous' means fiery, so it is not surprising that **igneous rocks** are formed from volcanoes. As the **lava** or molten rock comes out of the volcano it cools and forms rock. Lava cools quickly on the surface to form rock which contains small crystals. **Magma** cools slowly below the surface to form rock which contains large crystals.

A thin section of granite as seen through a microscope.

A thin section of porphyry as seen through a microscope.

### try this

**COLLECT**
samples of scoria, granite, pumice, gabbro, basalt, porphyry, rhyolite

### Examining igneous rocks

1. Draw up a table with the following column headings:
   Large crystals easily seen, Large and small crystals, Small crystals hard to see, Light colour or speckled, Dark colour or black.
2. Classify the rock samples under these headings.
3. Draw a key to classify these igneous rocks.

### your turn

1. Describe the major features of igneous rocks.
2. Explain how igneous rocks are different from sedimentary rocks.
3. Explain why igneous rock is difficult to break into pieces.

## Metamorphic rocks

When a caterpillar changes into a butterfly, this process is known as a **metamorphosis**. In the same way, **metamorphic rocks** are rocks that have been changed by heat, pressure and/or stress, as when mountains are formed.

Igneous rocks can turn into gemstones; for example, gabbro magma is rich in iron and magnesium and can form peridot, zircon, sapphire and diamond. Sedimentary rock can also be changed; limestone changes into marble and can also recrystallise to form rubies.

### try this

### A chemical circle

**COLLECT**
- limestone
- straw
- dropper
- filter funnel and paper
- glass beaker
- Bunsen burner
- heat-proof mat
- tripod
- gauze mat
- flat-bottom flask
- safety glasses
- tongs

Limestone is a sedimentary rock mostly made from the shells of tiny sea creatures that died a very long time ago. Limestone contains the compound calcium carbonate which can be changed by several chemical reactions.

**Caution:** Safety glasses must be worn for the entire experiment. Limewater is very dangerous to the eyes.

#### Chemical reaction 1

1. Heat the limestone piece very strongly on a wire gauze for several minutes. If you heat it strongly enough it will glow.

#### Chemical reaction 2

2. Leave the rock to cool. This rock is called lime.
3. Put the lime in the beaker. Add several drops of water. Listen carefully. Touch the bottom of the beaker.

#### Chemical reaction 3

4. Add about 20 mL of water to the lime. Filter the solution into a flask. The solution is called limewater.
5. Bubble your breath into the solution through a straw. Watch carefully. The carbon dioxide in your breath reacts with the limewater. The new substance that forms is calcium carbonate.

1. Draw a flow diagram to show the chemical changes that took place in this experiment, and the different substances that were formed.
2. Explain how this is an example of the processes that form metamorphic rocks.

### new words

**sediment • sedimentary rocks • igneous rocks • lava • magma • metamorphosis • metamorphic rocks**

# The formation of rocks  13.2

| NP | CSF |
|---|---|
| 4 | |
| 5 | |
| 6 | |
| 7 | |

The rocks you can pick up on a beach, on a mountain or in the garden are all made up of a combination of chemical compounds called **minerals**. The rock cycle below shows how new rock is formed over long time periods by processes on the Earth's surface and deep within the Earth.

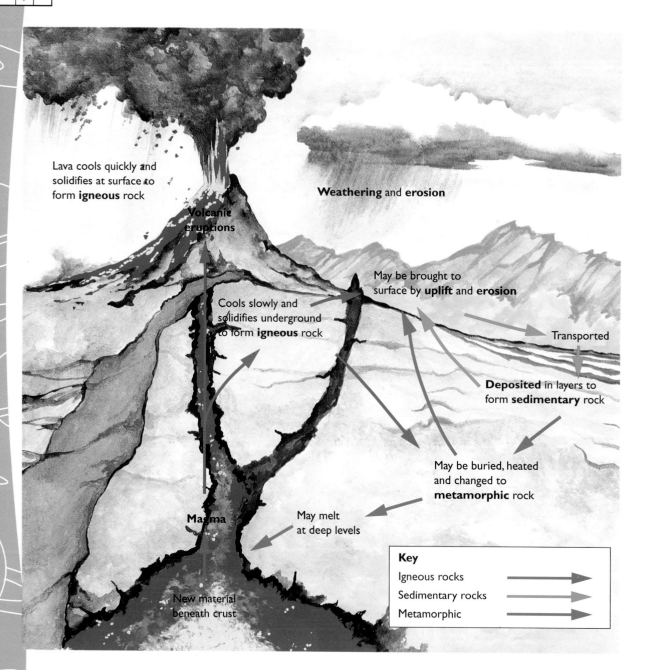

1  From the diagram, write a summary about how the different types of rock are formed.

### try this

### Making rocks (1)

The following activities use models to show how rock types form.

#### Sedimentary rocks

1. Add water to jar until it is two-thirds full. Measure the depth of water.
2. Pour the sample of different soils into the jar and swirl it gently. Allow the soils to settle.
3. Measure the new depth of water and the depth of the sediment.
4. Repeat the measurements next lesson.

**COLLECT**
- large jar
- ruler
- stirring rod
- sample of different sediments

#### Igneous rocks

When molten naphthalene cools it crystallises. Investigate the effect that different rates of cooling have on the appearance of the naphthalene crystals.

**Caution:** Be careful when heating napthalene – the fumes can ignite. You may need to do this experiment in a fume cupboard.

**COLLECT**
- cold and warm microscope slides
- hand lens
- pipette
- naphthalene

1. What are rocks made from?
2. Write a short report on each experiment. Remember to write a conclusion.
3. What rock would be produced from each of your sediments?

Your teacher may give you a rock cycle diagram to complete.

### new words

mineral • uplift • erosion • deposit • weathering

# project

# The uses of rocks

We use rocks every day, in gardens, for houses and other buildings, and to make other materials. Your task is to list as many rocks as possible and their uses. You will be assessed on how many rocks and uses you have listed, and on the thoroughness of your research. You must explain how the structure of your rock suits it for its uses.

Your teacher may give you an assessment criteria sheet.

**Look beneath your feet** 173

# Minerals and hardness 13.3

| NP | CSF |
|---|---|
| 3 | |
| 4 | |
| 5 | |
| 6 | |

Rocks are mixtures of different particles, including minerals. A mineral is a non-living substance that has a definite chemical content. It can be any part of the Earth's crust that is useful to us. Minerals are made from elements. For example, coal is made from the element carbon; salt is made from sodium and chlorine; rubies are made from aluminium, oxygen and chromium.

Minerals are identified using the following features:
- colour – the main colour
- **lustre** – the shininess of the surface, described as: dull, **vitreous** (glassy), brilliant, **pearly** or metallic
- **streak** – the colour the mineral leaves behind when it scratches another surface
- hardness – based on **Mohs' hardness scale**.

In 1822, the German mineralogist Friedrich Mohs devised a hardness scale using minerals as the points on his scale. Mohs' scale of hardness is shown below.

Talc.

Diamonds.

| Mineral | Hardness |
|---|---|
| Talc | 1 |
| Gypsum | 2 |
| Calcite | 3 |
| Fluorite | 4 |
| Apatite | 5 |
| Feldspar | 6 |
| Quartz | 7 |
| Topaz | 8 |
| Corundum | 9 |
| Diamond | 10 |

A hard mineral (high hardness number) will scratch a soft mineral (low hardness number). A mineral that scratches apatite but does not scratch feldspar has hardness between 5 and 6 – its hardness is said to be 5.5.

### on your own

**Cleavage and fracture: how the mineral breaks**

1. Choose one mineral to look at (e.g. talc, gypsum, mica, feldspar, quartz, malachite, azurite, bauxite, calcite, magnetite, galena, pyrite, limonite, sphalerite, haematite). Describe it and state what it is used for.
2. Find an example of a mineral and bring it to class. As a class, make a key for the minerals you have collected.

### new words

**lustre • vitreous • pearly • streak • Mohs' hardness scale**

# Rock for the road       13.4

The top layer of a sealed road is made up of pieces of rock bonded together by **bitumen**. However, not every rock is suitable for road making. The road surface should:
- be hard
- weather slowly
- prevent water from passing through it
- prevent cars from skidding.

To meet these requirements the rock selected must have the following properties.

**a** Most of the minerals in the rock should have a hardness greater than 5 on the hardness scale.

**b** The rock fragments must wear unevenly to make a skid-free surface. At least two different minerals of different hardness should be present.

**c** The rock fragments should have a rough surface for the bitumen to stick to.

**d** The rock must not let water through, nor shatter in cold weather.

**e** The rock should not be damaged by acid.

| NP | CSF |
|---|---|
|  | 4 |
|  | 5 |
|  | 6 |
|  | 7 |

### Rock information

| Rock | Minerals present |
|---|---|
| basalt | feldspar, augite |
| dolerite | feldspar, augite |
| flint | quartz |
| gabbro | feldspar, augite |
| gneiss | quartz, feldspar, mica, hornblende |
| granite | feldspar, quartz, mica |
| limestone | calcite |
| marble | calcite |
| sandstone | quartz |
| shale | clay minerals |
| slate | mica |

**try this**

## Properties of rocks

❶ Investigate the properties of the rock samples. Make up a checklist for your tests. Compare the rocks by:
- using available information
- making observations
- designing and carrying out tests for properties **d** and **e** in the list above
- estimating their hardness using the tests below.

**COLLECT**
- samples of feldspar, augite, quartz, mica, hornblende, calcite
- hand lens
- nail
- glass slide
- steel blade
- metal file
- bottle of acid
- dropper
- safety glasses

1. fingernail crushes
2. fingernail scratches
3. iron nail scratches
4. glass scratches
5. steel blade scratches
6. metal file scratches

**1** Write a report for the civil engineer in charge of building a road, describing the tests you did. Recommend one of the rocks for the road's surface, giving reasons for your decision.

# 13.4 continued...

## More rocks

Rocks normally take millions of years to form. It is easy to make some artificial 'rocks' that share the same features of true rocks.

**try this**

**COLLECT**
- dry clay
- dry sand
- plaster of Paris
- pebbles
- mortar and pestle
- beaker
- nylon stocking
- rubber band
- large glass jar
- moulds (foil-lined soft drink or fruit juice packs make good moulds)
- 2 blocks of shale
- tripod
- pipe clay triangle
- Bunsen burner
- evaporating dish

### Making rocks (2)

#### Sedimentary rocks

① Grind two tablespoons of dried clay-rich mud. Sift this as shown in the diagram. Use this clay to make your sedimentary rocks.

| Shale | Sandstone | Conglomerate |
|---|---|---|
| 5 spoons dry clay | 4 spoons coarse dry sand | 4 spoons of small pebbles |
| ½ spoon dry sand | ½ spoon dry clay | ½ spoon dry clay |
| 2 spoons water | ½ spoon plaster of Paris | ½ spoon plaster of Paris |
|  | 2 spoons water | 1 spoon coarse dry sand |
|  |  | 2 spoons water |

② Mix the dry ingredients on a sheet of paper, scrape into a heap. Make a shallow hole in the top of the heap, add half the water, mix until moist.

③ Press the mixture into a mould, sprinkle in the rest of the water.

④ Allow to dry and set (two to three days).

#### Metamorphic rocks

① Dry two blocks of shale for at least one week.

② Heat one block strongly as shown, for 30 minutes.

③ Cool the block and drop both into water. Which block crumbles? Explain your observations.

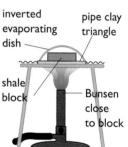

1 Write a report on your rock making. Describe what you did and your results.

2 Compare the properties of your rocks with samples of real rocks.

3 Present your results in a poster.

4 Investigate how **fossil fuels** are formed, using any of the resources available to you.

**new words**

bitumen • fossil fuels

# Coal 13.5

Coal is used to make many products that we take for granted: plastics, explosives, ammonia, medicines, nylon, cosmetics, fertilisers, bricks, tiles and soap are just a few. These products are made from the coal tar which is produced when coal is converted into coke. The coke itself is used in smelting iron ore to make steel.

Coal is also an important energy source. Steaming coal is used in power stations to create the electricity we use to run many appliances in our homes and businesses.

Coal is plant material that, over millions of years, has decayed and been subjected to heat and immense pressure. Scientists have estimated that a 120 metre thickness of plant material is needed to produce a 1-metre thickness of hard black coal. Layers of coal are called **seams** and large amounts of coal in one spot are called **deposits**. In Australia, our major deposits of coal are found in the Bowen Basin in Queensland, in the Sydney Basin in NSW and the Latrobe Valley in Victoria.

Most of the coal in the Bowen Basin is extracted using **open-cut mining**. First the coal seam has to be exposed. This involves blasting and drilling the top layers of rock and soil and carting this away. This material is called **overburden**. Once the coal is exposed it is also loosened by blasting and then carried away to the preparation plant. Here the rock, low-quality coal and impurities are removed to produce high quality coking coal for steel manufacture.

Australian coal is sent to many overseas markets including Japan, Korea, India, Turkey and Mexico.

When an open-cut mine ceases production, the area must be returned to a useable state. These days mining companies collect detailed information on the type of plants and animals that live in an area for just this purpose. The overburden and topsoil are replaced and landscaped, and trees are planted. Many thousands of dollars are spent on **rehabilitating** the area.

| NP | CSF |
|---|---|
| | 4 |
| | 5 |
| 6 | |
| | 7 |

## your turn

1. Explain what coke and steaming coal are used for.
2. Explain whether we could live without coal.
3. Describe how coal is mined in Queensland. Draw a flow diagram of the mining process.
4. Explain how the environment is affected. Describe how the mine is left.
5. Explain whether the environmental costs of mining outweigh the benefits.

## new words

Add any new words to your glossary.

# another look  13.6

### 13.1 Three types of rocks
1. Describe the different types of rocks.
2. Give two examples of each type of rock.
3. Describe the differences between the three types of rocks and the features you would use to tell them apart.
4. Draw diagrams of each type of rock.

### 13.2 The formation of rocks
1. Choose three rocks mentioned in this section and state how these rocks are formed.
2. Explain how the different types of rocks are formed.
3. Describe how you would make sedimentary and igneous rocks.

### Project: The uses of rocks
1. List five rocks and their uses.

### 13.3 Minerals and hardness
1. Describe a mineral.
2. Explain how minerals are classified.
3. If you found a mineral that scratched feldspar but didn't scratch topaz, what mineral would it be?
4. Describe a mineral that you studied.

### 13.4 Rock for the road
1. List the necessary properties of rock used for road making.
2. Explain how you could test which rocks are best for road making.
3. Describe how you would make sedimentary and metamorphic rocks.

### 13.5 Science in Action: Coal
1. Briefly describe the importance of coal in our everyday lives.

# another look

## How rocks are formed

Copy the following flow chart into your workbook and complete it by adding the missing labels.

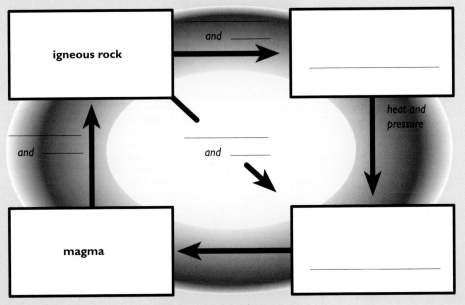

## Making a concept map

In groups, choose one of the following lists of words to use in making a concept map. Swap your map with another group and add as many details as you can.
a   rock, geologist, igneous, volcano, lava, heat, Earth, metamorphic
b   rock, change, metamorphic, heat, minerals, marble, Earth
c   sedimentary, rock, sand, water, time, fossils, sandstone, Earth, metamorphic.

## Linking concepts

Look back at the work you have done in chapters 12 and 13. Draw a concept map to show how these chapters link together.

## Strengths and weaknesses

Don't forget to identify your strengths and weaknesses in this section of work.

Your teacher may give you a summary sheet and a chapter test.

# Reading and analytical skills

Often we need to compare two pieces of writing. When you compare pieces of writing, read each piece separately and take notes on each passage.
- First, skim the passage to find the main ideas.
- Next, read the passage more carefully and make notes as you read.
- Repeat this procedure with the second passage.
- Put your notes under different headings for easy comparison.

## Passage A

The universe of the Egyptians was a more rectangular oyster or box; the Earth was its floor, the sky was either a cow whose feet rested on the four corners of the Earth, or a woman supporting herself on her elbows and knees; later, a vaulted metal lid. Around the inner walls of the box, on a kind of elevated gallery, flowed a river on which the Sun and Moon gods sailed their barques, entering and vanishing through various stage doors. The fixed stars were lamps, suspended from the vault, or carried by other gods. The planets sailed their own boats along canals originating in the Milky Way, the celestial twin of the Nile. Towards the fifteenth of each month, the Moon god was attacked by a ferocious sow, and devoured in a fortnight of agony; then he was reborn. Sometimes the sow swallowed him whole, causing a lunar eclipse; sometimes a serpent swallowed the Sun, causing a solar eclipse. But these tragedies were, like those in a dream, both real and not; inside his box or womb, the dreamer felt fairly safe.

From *The Sleepwalkers*, by A. Koestler (Hutchinson 1968)

## Passage B

We approach the planets of our system, largish worlds, captives of the Sun, gravitationally constrained to follow nearly circular orbits, heated mainly by sunlight. Pluto, covered with methane ice and accompanied by its solitary giant Moon Charon, is illuminated by a distant Sun, which appears as no more than a bright point of light in a pitch-black sky. The giant gas worlds, Neptune, Uranus, Saturn – the jewel of the solar system – and Jupiter all have an entourage of icy moons. Interior to the region of gassy planets and orbiting icebergs are the warm, rocky provinces of the inner solar system. There is, for example, the red planet Mars, with soaring volcanoes, great rift valleys, enormous planet-wide sandstorms, and, just possibly, some simple forms of life. All the planets orbit the Sun, the nearest star, an inferno of hydrogen and helium gas engaged in thermonuclear reactions, flooding the solar system with light.

Finally, at the end of all our wanderings, we return to our tiny, fragile, blue-white world, lost in a cosmic ocean vast beyond our most courageous imaginings. It is a world among an immensity of others. It may be significant only for us. The Earth is our home, our parent. Our kind of life arose and evolved here. The human species is coming of age here.

From *Cosmos*, by C. Sagan (Macdonald Futura 1981)

1 According to the theory in passage A:
   a What was at the centre of the heavens?
   b What caused the phases of the Moon?
   c What caused a solar eclipse?
   d What are the stars?
   e How do the planets move?

2 According to passage B:
   a What is at the centre of the solar system?
   b What causes the planets to follow their orbits?
   c What causes the Sun to give out energy?
   d What are the stars
   e How do the planets move?

# index

**A**
Aboriginal beliefs 157
accuracy 7
acid, dangers of 61
acids 55–65
air sacs 72
airway 70
alloy 54
arteries 74–7
atoms 28, 30, 32, 38, 41, 49

**B**
bases 55–6
bends, the 78
bibliographies 13
bleeding 77
blood pressure 77
boiling point 29
burns 16, 40, 77

**C**
cancer 92
capillaries 74
carnivores 102
cell division 89
cell reproduction 89
cell specialisation 90
cell wall 87
cells 82, 86–94
    parts of 87
chemical change *see* chemical reaction
chemical energy 116
chemical reaction 40, 47, 48
chlorine 50
chloroplasts 87
circulation 73–7
circulatory system 75
coal 177
communities 100
compounds 28, 42–4
computers 134
conditions (environmental) 96, 97
conduction 118–20
consumers 100
convection 118–20
cranial nerve 20

**D**
danger 69
deafness 150–1
decibels 149
decomposers 100
diaphragm 71
diffusion 88
divers 78
DR ABC 69–73
ductility 34

**E**
ear, parts of 21, 144
Earth
    orbit 158
    resources 36
    revolution 158
    rotation 158
eclipses 161–2
ecosystems 100–3
elements 28–35, 40
    appearance and properties of 32
    in the human body 31
energy 114–22, 127–33, 138–41
    forms of 114–15
    transfer of 114, 118, 120
    transformation of 114–17
eye 17–19, 61
eyepiece lens 85

**F**
fire extinguishers 60
first aid 69–78
food chains 100–2
food webs 103
formulae (of molecules) 41–9
fractures 77
frequency (of sound waves) 142
fulcrum 130

**G**
gears 129
gravity 159, 161
ground water 106

**H**
habitat 96–103
hearing range 142
herbivores 102
hydrogen 28, 30, 31, 38, 40, 41, 54, 55, 56, 59
hypothesis 3, 7–8, 9, 10

**I**
igneous rocks 169–70, 172–3
inclined planes 128
indicators 55–8
infra-red radiation 120
Internet 134
irrigation 106

**K**
kilohertz 142

**L**
lateral recovery position 70
lava 170
Leaning Tower of Pisa 10
levers 130–1
longitudinal waves 140
lungs 71–2

**M**
machines 126–34
    complex 132, 133
    simple 128–32
magma 170
magnification 85
malleability 34
melanomas 22
melting point 29
metabolism 89
metals 31–6, 46, 54, 145
metamorphic rocks 169, 171–2, 176
microscope 68, 82–6, 90
    binocular and monocular 83
minerals 172, 174
mitochondria 89
Mohs' hardness scale 174
Moon 157, 158–62
    phases of 159
    walking on 160

**N**
neutral substances 55, 58
neutralisation 59
noise 151–8
nuclear energy 116
nucleus 87
nutrients 104

**O**
objective lens 85

olfactory membrane 20
omnivores 102
orbit (of the Earth) 158
organs 91
oscilloscope 140
ozone 22

**P**
periodic table 30
pH paper 57
pH scale 57
photosynthesis 100
pitch 142
pivot 130
place 96
plasma 74
platelets 74
populations 98
potential energy 116
producers 100
pulley 128
pulse 76

**Q**
qualitative data 9
quantitative data 9

**R**
radiation (heat) 120
ramp 128
receptors 20
recycling of matter 104–5
red blood cells 74

resonance 148
respiration 89
response 69
rocks 169–76
    formation of 172–3
    properties of 175–6
    types of 169–71
    uses of 173, 175–6
rotation (of the Earth) 158
rust 45–6

**S**
salinity 106
saliva 20
salts 59
scientific method 2
screw 128
sediment 169
sedimentary rocks 169
sight 17–19
    long- and short-
      sightedness 18
silicon 122
skin 14
skin cancer 14, 22
solar cells 122
sound energy 139
sound intensity 148–9
sound waves 140–2
    longitudinal 140
    transverse 140
space shuttle 164
specialisation (of cells) 90–1

speed of sound 146
symbols (of elements) 30–1, 49

**T**
taste buds 20
thermal pollution 107
tides 161–2
timbre 148
tissue 91
tone (of a sound) 148
trachea 71
transverse waves 140
tumours 92

**U**
unconsciousness 69
universal indicator 57
UV light 22

**V**
vacuoles 87
valves 74
variables 4–6
veins 74
vibrations (sound) 139–42
vocal cords 144

**W**
wheels (and axles) 129
white blood cells 74
windpipe 71
work 127